Controlling Air Movement

Controlling Air Movement

A Manual for Architects and Builders

Terry S. Boutet

McGraw-Hill Book Company

New York   St. Louis   San Francisco   Auckland   Bogotá
Hamburg   London   Madrid   Mexico   Milan
Montreal   New Delhi   Panama   Paris   São Paulo
Singapore   Sydney   Tokyo   Toronto

Library of Congress Cataloging-in-Publication Data

Boutet, Terry S.
  Controlling air movement.

  Bibliography: p.
  Includes index.
  1. Ventilation. 2. Air conditioning. I. Title.
TH7653.B68 1987   697.9′2   87-3163
ISBN 0-07-006713-9

1234567890 DOC/DOC 893210987

ISBN 0-07-006713-9

$$\begin{array}{c} D \\ 697.92 \\ BOU \end{array}$$

*The editors for this book were Nadine M. Post
and Galen H. Fleck, the designer was Naomi Auerbach,
and the production supervisor was Richard A. Ausburn.
The illustrations were by Terry S. Boutet.
This book was set in Century Schoolbook. It was composed by the McGraw-Hill Book
Company Professional & Reference Division composition unit.*

Printed and bound by R. R. Donnelley & Sons Company.

To my supportive and loving wife, Carol, and our daughter, Jocelin.

Contents

Preface

Modern architectural practice embraces a wide field of information from building materials to behavioral science; consequently, reference books are the greatest tools available to conscientious designers. *Controlling Air Movement* is such a tool. It is a guide to air movement in residential structures and utilizes over 300 illustrations. Its objective is to present the variety of design options for air movement control that are available to designers and the impact those options have upon the final building form, total design image, solid-void relationships, and the structure's relationships to its site configuration, orientation, topography, and landscaping.

This book centers on implementing the positive effects of air movement as a way to improve the quality of living through optimum application to residential structures. The benefits include improved air quality, energy conservation with reduced operational costs, and superior human physical and mental comfort.

In addition, methods of studying air movement are compared and analyzed. Basic principles of air movement control are thoroughly described with an emphasis on building forms, openings, projections, and partitions.

From the data presented in this book, a thorough index, described and illustrated, of air movement control techniques is developed. Those techniques, tested with models in a kerosene smoke airflow chamber, consisted of designs by practicing architects and the author. Consistently, the focus of *Controlling Air Movement* is on the potential benefits of securing optimum air movement in residential structures through the application of proper control techniques.

To date, no single presentation of information on air movement control has covered all the possible site characteristics, building configurations, opening variations, and other design modifications, nor has any presentation documented the various probabilities. I have personally searched through thousands of books and articles in the engineering, architectural, medical, agricultural, and energy fields without finding a single document that begins to contain the volume of information found in this book. Also, the format permits easy usage and quick reference.

Architectural designers cannot spend great amounts of energy and time resolving the climatic needs of each and every design problem, nor can they start with assumptions about important features of each building with regard to air movement control. Instead, they need ready facts upon which to base intelligent decisions. This book unites the explorations of past researchers and develops a relatively comprehensive guide of known air movement control techniques. In short, this book is a necessary tool for conscientious designers whether they are professional architects, architectural educators, intern architects, or architectural students.

Controlling Air Movement is a valuable reference for the professional architectural society as well as other building designers, and it will provide a needed service and be a wise investment for architectural designers. designers.

Terry S. Boutet

Chapter

1

Air Movement

Architectural designers can no longer ignore the effects of climate on buildings; they must understand and use the fundamentals of air movement to improve their designs. Though the task is complex, they must consider the prevailing weather conditions to build durable, efficient, aesthetic, and economical structures.

Climate affects every design solution however sophisticated; prevailing weather conditions alter the environment in and around structures. Thermal and moisture distribution systems are the major factors, and thermal distribution has the greater effect on comfort. Since air movement distributes most of the earth's heat, architectural designers must use it to design for comfort. To do so requires a thorough understanding of basic principles.

Terminology

Some terms need clarifying; others will be defined as they are introduced. Various forces cause air to move across the earth. The types of air motion are defined by those forces.

Air, the atmosphere that surrounds the earth, is a mixture of gases. *Air movement* is a change in position of air regardless of the cause or degree.

Wind is the natural form of air movement; usually but not always the movement is horizontal.

Ventilation is the process of supplying unconditioned or conditioned air to and removing it from a given space by any method. For example, when wind enters a bedroom, it becomes ventilation, but it is always air movement.

Ventilation is classified by the force acting on the air: *Natural venti-lation* depends on natural forces; *induced ventilation* depends on influencing natural forces to perform specific tasks as in a thermal chimney; *forced ventilation* depends on mechanical methods. *Cross-ven-tilation,* a loosely applied term, actually refers to air movement across a space connected by openings to both positive and negative pressure areas of the exterior.

Buoyancy, a force that creates air movement, is the tendency of air to rise because of thermal differentials (Figure 1.1). It is *positive, neutral,* or *negative* in reference to its ability to move air. High air density creates

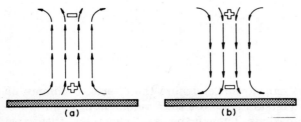

Figure 1.1   Variations in air density due to thermal differentials may cause air to (*a*) rise (buoyancy) or (*b*) fall (reverse buoyancy).

positive buoyancy; low air density creates negative buoyancy. Air moves from positive to negative through neutral buoyancy, in which the air density is neither high nor low. *Reverse buoyancy* may occur when the air tends to fall because of thermal differentials. Whatever the cause of buoyancy, air movement must be from positive to negative.

Pressure, a force that creates air movement, is the shifting of air caused by force differentials (Figure 1.2). It also is positive, neutral, or negative. Positive pressure occurs when the force and air density are high; negative

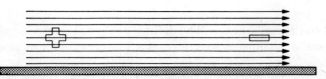

Figure 1.2   The action of force differentials on air movement is known as pressure. Air moves from positive to negative pressure zones.

pressure occurs when the force and air density are low. Air travels to nega-tive from positive through neutral pressure, in which the force and air density are neither high nor low. Air movement always occurs in a posi-tive-to-negative fashion.

Climate, the prevailing weather conditions of a given region, is commonly considered at three principal levels (Table 1.1). The *macro-climate* comprises the weather conditions of a large region such as a continent or country. The *meso-climate* comprises the weather conditions of a region that is neither large nor microscopic, such as a state, county, or city. The *microclimate* comprises the weather conditions of a very small, or microscopic, region such as a leaf, blade of grass, or sidewalk crack. A relatively new term is *micro-climate*, which is easily confused with microclimate. A micro-climate comprises the weather conditions of a small region such as a city block, acre, lot, or garden.

TABLE 1.1   Principal Climatic Scales

Climate	Scale	Example
Macro-climate	Extremely large	Continent Country
Meso-climate	Moderate	State County City
Microclimate	Small	City block Acre Lot Garden
Micro-climate	Extremely small	Leaf Blade of grass Sidewalk crack

Air Movement Systems

Air movement is created by uneven heating of the atmosphere. As the sun heats it, the air expands, rises, and is replaced by cooler air. The exchange of air creates a cycle known as the *general circulation*, the major wind system of the earth.

Occurring as it does over large portions of the earth, the general circulation creates prevailing air movement. It is affected by differences in heating qualities of land and sea and by the position of the sun. Land is heated by the sun quicker than water; water retains more heat for longer periods of time. Consequently, the land and water almost always differ in temperature, and that causes the air masses over them to move (Figure 1.3). The sun constantly changes its position, and so the angle of

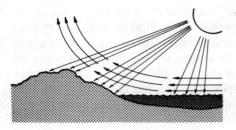

Figure 1.3 Temperature differences between land and sea cause air masses over land and sea to move. As the air rises over the warmer body, cooler air replaces it.

its rays varies through the year. The effect of the variation is to heat the earth unevenly and at varying rates. Because of that uneven heating, there are three main global belts of the general circulation in each hemisphere (Figure 1.4).

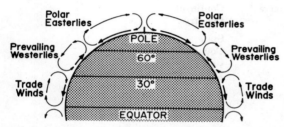

Figure 1.4 The prevailing air movement of the general circulation occurs in three main global belts within each hemisphere. The heavy arrows indicate the direction of air movement across the face of the earth.

The *trade winds* travel to the equator from approximately the 30° latitude in both hemispheres; in those regions the air moves from positive pressure to the equator's negative pressure zone. At the equator, the air rises, returns to the 30° latitude zone, and thereby creates a cycle.

The *prevailing westerlies* abut the trade wind belts in both hemispheres. Their cycles are opposite to those of the trade winds; the air moves to the negative pressure zones at 60° latitude from the positive pressure areas at 30° latitude.

The third global belt is the *polar easterlies.* The air moves from the north or south pole to the 60° latitude zone in a positive-to-negative pressure fashion. Where global belts abut, some air moves from one belt to another.

The global belts do not move in direct north-to-south or south-to-north fashion, however; instead they are curved. The cause of the curving, known as the Coriolis force, is the earth's rotating faster than the atmosphere, which is slowed by the friction of the earth's surface (Figure 1.5).

The force is zero at the equator, increases toward the poles, and has a magnitude proportional to the sine of the latitude.[1]*

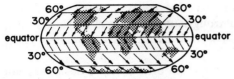

Figure 1.5 The earth's rotation creates a curving action of the global belts called the Coriolis force.

Secondary Circulation Systems

Within the general circulation system are several air movement patterns that have lesser intensities. They occur over small areas that usually involve positive and negative pressures (Figure 1.6). The air movement from positive to negative pressure areas is circular. Positive pressure areas are mountains of air as they rise above surrounding air because of increased air density, and the air movement spirals downward and outward. In negative pressure areas, the valleys of low air density, air movement occurs in a downward and inward spiral and causes the air in the center to rise. In these negative pressure areas, the rising air cools and condenses and often creates clouds, fog, and rain. These secondary circulation systems are unaffected by minor variations of the earth's surfaces, but they are affected by major land forms of the earth such as plains, mountains, valleys, plateaus, and seas.

Figure 1.6 Air pressures of the secondary circulation systems vary as the air moves from positive to negative pressure areas.

Meso-climatic Systems

When the secondary circulation systems develop variations within themselves that are due to cities, housing developments, industrial centers, and so forth, meso-climatic systems are established. The driving forces at this

·* Numbers in brackets are those of references listed at the ends of chapters.

level are pressure and buoyancy, which may work together or independently. Although their intensities vary, buoyancy is the weaker of the two,
since thermal differentials are not as easily obtained as pressure differences. Buoyancy usually creates vertical air movement, whereas pressure
creates horizontal air movement. Throughout the meso-climatic systems,
a great variety of air movement may occur because of pressure, buoyancy,
and terrain.

It is within the meso-climatic systems that architectural designers can
begin to utilize air movement by controlling and altering the driving
forces. On the micro-climatic scale, air movement can be transformed into
ventilation traveling through architectural designs. A wealth of free
energy is waiting to be used by the architectural designer.

REFERENCES

1. B. Givoni, *Man, Climate, and Architecture*, Applied Science, Ltd., England, 1976.

2

Effects of Air Movement

Air movement, which influences air purity, temperature, and moisture, has a direct effect on human health, comfort, and well-being. Although occupant well-being is of primary concern, the building envelope must be considered as well. The way a residence relates to its microclimate determines the type of climatic environment within the structure. Consequently, air movement plays an active role in the quality of life.

Air movement in residences has three separate functions: *air quality*, *energy*, and *comfort*. B. Givoni called them "health ventilation," which refers to maintaining the air quality by replacing indoor air with fresh outdoor air, "structural cooling ventilation," which describes the method of cooling the structure when the indoor air temperature is higher than the outdoor air temperature, and "thermal comfort ventilation," which provides for a reduction of heat and moisture from the body into the surrounding air. [1]Although Givoni's categories are limited in scope, they are a basis for understanding the benefits of air movement with buildings.

Air quality, energy, and comfort describe the functions more thoroughly and are used in this book. *Air quality,* which describes the characteristics of the air, is a more suitable term than health ventilation. Not all health problems are caused by lack of air purity. The architectural designer must be concerned with both interior and exterior spaces. Ventilation starts where air enters the building, but its source of energy is the air movement of larger and more encompassing systems.

Energy involves both avoidance of heat gain and acceleration of heat loss, whereas Givoni's structural cooling employs only the release of gained heat.

Comfort encompasses both the physical and psychological aspects of human well-being. Givoni's thermal comfort, however, involves only the physical aspect.

Air Quality

Pollution arising from fuels used in homes to cook or create comfort and from human respiratory processes has always plagued humanity. Patricians of the Roman Empire complained about togas soiled by soot from "the heavy air of Rome and the stench of its smoky chimneys."[2] Air pollution, in residences as well as factories and plants, remains an important issue, and it is becoming more involved as technology advances. The solution is to stop air pollution at its source or to dilute pollutants so that the air can be cleansed naturally.

One of the easiest places to control residential air pollution is at its source within the home. Control must begin before the building is occupied. Many finishing materials used in construction, especially residential, introduce pollutants. Over 70 percent of a home's interior consists of materials that contain formaldehyde, which is suspected to be a human carcinogen (Figure 2.1). Even a concentration as low as one part in 100 million parts of air makes eyes water. In sufficient quantities, formaldehyde can cause lung irritation, nausea, vomiting, drowsiness, sore throat, headache, and fatigue.[3]

Figure 2.1 Formaldehyde from carpet (1), paneling (2), particleboard cabinets (3), wallpaper (4), curtains (5), and plastics (6) poses potential health threats unless the home is adequately ventilated.

The problem is not just with new building materials; it may last as long as 20 to 30 years.[3] Homeowners add still more formaldehyde in a variety of products: plastic appliances, permanent-press sheets, towels, soaps, plastic trash bags, hair spray, newspapers, cosmetics, clothing, disinfectants, plastic wraps, shampoo, and toothpaste. Nevertheless, the trouble can be reduced or eliminated by utilizing building materials with low levels of formaldehyde, sealing the materials with a water vapor barrier, and providing adequate ventilation.

Radon, a byproduct of the radioactive decay of radium, cannot be avoided. It is found in soils and rocks, most commonly in Texas, Pennsylvania, Tennessee, Colorado, New Hampshire, Maine, Montana, and Florida and especially in residences built over old granite and phosphate mines (Figure 2.2). Radon was found in excessive levels in 15 percent of thousands of homes tested in eastern Pennsylvania, posing a high lung cancer risk.[3] "Indeed, health authorities now believe that exposure to radon is the second leading cause of lung cancer in the United States, accounting for some 5000 to 20,000 new cases per year."[4] Unfortunately, it cannot be completely eliminated, but its quantity can be reduced by as much as 90 percent by isolating the building from the soil and providing proper ventilation.

Figure 2.2 Radon, a radioactive byproduct found in concrete (1), sand (2), gravel (3), stone (4), brick (5), soil (6), and drinking water (7), may become confined within a closed house and concentrated to levels up to 10 times higher than those outdoors.

Household products introduce other interior pollutants. "In the home of Mr. and Mrs. Average American are 45 aerosol sprays."[3] Examining the home reveals a large number of other products that may cause pollution (Figure 2.3). A scented product adds substances to the air that may or may not be desirable. Furthermore, the problem of household product pollution is dual: pollution from the product's chemical content may

	Acid	Benzene	Chloride	Formaldehyde	Hydrocarbons	Hydroxides	Lye	Nitrous oxides
Aerosol sprays					▓			▓
Bath soaps				▓		▓	▓	
Deodorants			▓		▓		▓	
Detergents				▓		▓	▓	
Disinfectants	▓		▓	▓				
Drain cleaners								
Hair sprays	▓							
Moth balls		▓	▓		▓			
Typical Bathroom	▓	▓	▓	▓	▓	▓	▓	▓
Typical Kitchen	▓		▓	▓	▓	▓	▓	▓

Figure 2.3 Household products pose a definite threat of headaches, rashes, burns, nausea, depression, and rapid heartbeat as they introduce a variety of chemicals into the typical home. *(Data derived from Ref. 3)*

be harmful, and the pollutants in two or more products can have a who-knows-what chemical reaction. Household product pollution can be alleviated by proper selection and use of the products and by adequate ventilation.

People are the final major source of interior air pollution; metabolism produces byproducts discharged in respiration. Outdoors they are easily removed by air movement; indoors they become concentrated. Respiratory byproducts include carbon dioxide discharged from the lungs, bacteria expelled with the breath, and odor given off by the body. Fluctuations of oxygen and carbon dioxide content in conventional residences have little significance. Physiological studies show that "the concentrations of carbon dioxide and oxygen are not suitable as direct criteria for the specification of ventilation requirements." [1]

Unlike carbon dioxide, which may be influenced by ventilation, bacteria remain at a constant level and are unaffected by increased ventilation.[1] Consequently, bacterial growth is affected only by certain chemicals and ultraviolet light. Body odors are unstable and will usually dissipate on their own from an objectionable to a perceptible level. However, the requirement should be that disagreeable odors cannot be detected in a residence. Like carbon dioxide, body odor concentrations are not high enough to be suitable criteria for ventilation specification. Consequently, adequate ventilation to dissipate chemical odors, such as household products, is more than sufficient to remove respiratory byproducts.

Exterior air pollution is not easily controlled; the source is often unknown. A simple way to eliminate exterior air pollution is to avoid it by selecting a residential site windward of a known source of pollution, such as a factory, or in an area known to be pollution-free. If air pollution cannot be avoided, filter the air. The viscous surfaces of plant leaves capture dust and remove particulates from the air. Vegetation cannot remove gaseous air pollution, but most air pollutants are filtered by vegetation or washed down by rain. Side effects, such as withering trees, dying lakes, and corroding buildings, will occur if nature's laundry system is overloaded with wastes.

Energy

Proper air movement control lessens the demand for energy, thus reducing the expense of providing a comfortable home. Architectural designers can reduce direct heat loads to cool the home environment. Heat gain from the sun can be avoided by using shading techniques; windows in particular add significantly to residential heat gain. Ventilation of the attic space reduces major heat gain, since the sun's rays impact mainly on the roof. Avoiding reflective surfaces around the building prevents retransmission of solar heat into the home. Mechanical cooling systems protected from direct sunlight function with less mechanical stress and greater efficiency. Residents can reduce interior heat loads by turning off unneeded lights and using heat-producing appliances, such as dishwashers, during the coolest part of the day.

Designers can also increase the heat loss as a way of cooling the building and its environment; two-thirds of the heat loss is by conduction through the roof, walls, and windows.[5] Exposing more of the building to air movement by increasing the surface area and getting greater surface-air contact may provide more structural cooling. Heat loss occurs by turbulent mixing of air close to the wall surface, turbulence inherent in the air flow, air movement patterns around the building, and turbulence generated in the wall boundary layer.[5]

Comfort

Comfort and health have been, and always will be, influenced by climate. Hippocrates, in about 400 B.C., wrote that:

Whoever would study medicine aright must learn of the following subjects. First he must consider the effect of each of the seasons of the year and the difference between them. Secondly, he must study the warm and the cold winds, both those which are common to every country and those peculiar to a particular locality. Lastly, the effect of water on the health must not be forgotten.

Thus he would know what changes to expect in the weather and, not only would he enjoy good health himself for the most part, but he would be very successful in the practice of medicine. If it should be thought that this is more of the business of the meteorologist, then learn that astronomy plays a very important part in medicine since the changes of the seasons produce changes in the mechanism of the body.[2]

Climate and health are still vital issues among modern doctors, as Sir Leonard Hill pointed out in a report to the Medical Research Council:

The changing play of light, of cold and warmth, stimulate the activity and health of the mind and body. Monotony of occupation and external conditions for long hours destroy vigour and happiness of, and bring about the atrophy of disuse in, men.[2]

Climate significantly influences health and well-being and sets a foundation on which to establish human comfort, the ideal conditions for a person to perform at optimal level with supreme physical and psychological fitness.

Comfort is derived from many physical and psychological factors that produce a state of satisfaction or make life easier, and not from thermal comfort alone. Thermal comfort is the condition which produces minimal activity of the thermoregulatory mechanisms of the body. Comfort, rather than being a set standard, fluctuates with the factors that produce it. The *comfort zone* is defined as the range of conditions in which more than 50 percent of the persons tested felt comfortable (Figure 2.4). Although test data were based on physical measurements, psychological factors were built into the test because the subjects were giving a biased opinion with regard to comfort. The physical aspects of comfort depend on six major factors that function as an intertwining system influenced by psychological factors.

First, ambient air temperature provides a foundation for measuring comfort. According to William A. R. Thomson, a naked person finds exposure to still air at 99 to 95°F to be bearable, 80 to 77°F to be pleasant, 59°F to be cold, and 53°F to be extremely cold in a few minutes. [2]

Second, mean radiant temperature modifies the effect of the ambient air temperature. It is the effect of surrounding surface temperatures, which vary with space and time. To some extent, it may balance higher or lower ambient air temperatures, but its role is usually minor; in typical architectural practice the difference between air and surface temperatures is not more than 4 or 5°F.[7] In extreme situations, however it may be a significant factor. For example, a person standing by a large window

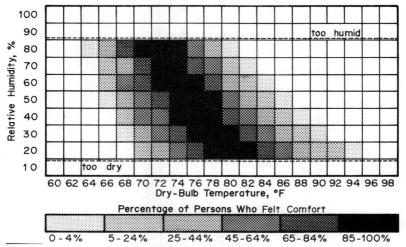

Figure 2.4   The chart illustrates the range of conditions which may provide comfort based on test studies performed by ASHRAE. *(Data derived from Ref. 6)*

during a hot Florida summer day may feel an increase in temperature of 4°F if the window is shaded and 8°F if it is not.

Third, humidity appears to have a more direct effect on comfort than the mean radiant temperature. Although humidity does not add to the body heat load, it affects the body's capacity to dissipate heat through evaporation. The body really feels the effect of humidity when the ambient air temperature is below 68 or above 77°F.

Fourth, air movement removes excessive heat by increasing convection and evaporation rates. The cooling rate increases with an increase in air velocity. When the ambient air temperature is below body temperature, a rise in air velocity always produces a cooling effect which increases as the air temperature decreases (Table 2.1). When the ambient air temperature is above body temperature, a rise in air velocity warms and cools the body at the same time. However, the cooling effect is greater than the warming effect until the air temperature reaches approximately 104°F, at which the warming effect is greater.[1]

Fifth, clothing affects the body's sensitivity to climatic variations because it interferes with evaporation and forms a barrier to convection (Figure 2.5). It also reduces the influences of lower ambient air temperatures and mean radiant temperatures. Lightweight pants, for example, let out 20 percent more body heat than jeans, and shorts permit 120 percent more heat to escape; a short-sleeved shirt doubles body heat loss compared to a long-sleeved shirt.[8]

TABLE 2.1 Wind Chill

Wind speed, mph	Wind chill when ambient air temperature, °F																
	40	35	30	25	20	15	10	5	0	-5	-10	-15	-20	-25	-30	-35	-40
5	38	32	28	21	16	12	6	1	-6	-10	-15	-20	-26	-31	-36	-41	-47
10	27	21	16	10	3	-3	-9	-12	-21	-27	-32	-39	-46	-52	-58	-64	-70
15	22	15	10	2	-5	-11	-18	-26	-34	-40	-45	-51	-59	-65	-71	-78	-85
20	18	11	4	-3	-10	-17	-25	-32	-40	-46	-52	-60	-68	-75	-81	-88	-96
25	15	7	0	-7	-15	-22	-29	-37	-44	-51	-59	-67	-74	-81	-87	-95	-104
30	12	5	-2	-11	-18	-26	-33	-40	-48	-56	-63	-70	-78	-86	-94	-101	-109
35	10	3	-4	-13	-20	-28	-35	-43	-51	-59	-66	-73	-82	-90	-98	-105	-113
40	9	2	-5	-14	-21	-29	-37	-45	-53	-61	-69	-76	-86	-93	-101	-109	-116

Wind chill is the effective temperature created by low ambient air temperature and wind speed. Once the air velocity reaches 40 mph, no further chilling really occurs.

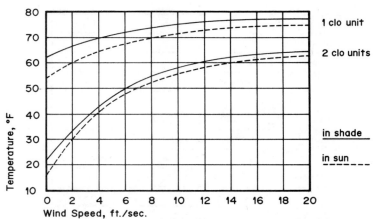

Figure 2.5 Several comfort curves for a pedestrian walking at 2.5 mph in an outdoor environment. The clo unit is a measurement of clothing's thermal insulation. Typical light clothing, office dress, equals 1 clo unit, and winter attire measures 2 clo units. *(Adapted from Ref. 5)*

Sixth, the metabolic rate of the body is the key to comfort. Heat losses that are too great cause freezing or death; heat gains that are too great cause stroke or death. The metabolic rate, which is proportional to the body's weight, increases with physical activity (Figure 2.6). The body needs more cooling as the metabolic rate increases and less cooling as the rate decreases.

Psychological elements also influence perception of comfort. People may feel comfortable if they think their environment provides comfort. A person may feel cooler on the front porch than in the kitchen because the shaded outside space is perceived as cooler when, in fact, both spaces provide the same level of comfort or the interior space may actually be more comfortable than the exterior one. Color influences psychological comfort. Light-colored textiles and throw rugs used in the summer create a light, airy, and cool atmosphere, and dark and rich colors used in the winter create a warm and cozy sensation when, in actuality, the thermal conditions remain the same.

Personal attitude can produce psychological comfort. A person who shuts off the air conditioner, opens the windows, and turns on the ceiling fans may feel more comfortable because of the saving in energy and money. Psychological comfort, one could say, is mind over matter—if you don't mind, it doesn't matter. But since psychological factors are not measurable, comfort must be measured by physical factors. Testing the physical factors of comfort on human subjects does insert psychological factors into the final results to some degree. Subjects include their

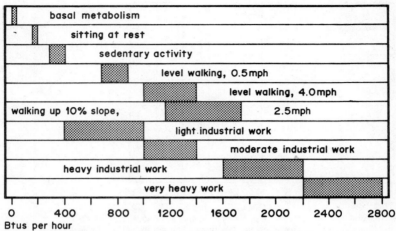

Figure 2.6 Metabolic rates for various activities. *(Adapted from Ref. 1)*

own biases—psychological factors—in their responses about whether
they are comfortable.

Human comfort has been studied by many; both Victor Olgyay and
Donald Watson derived individual comfort charts that have been utilized
in the past (Figures 2.7 and 2.8). A composite rearrangement of the two
charts creates a new Human Comfort Chart involving both physical and
psychological factors in a format readily usable by architectural designers
(Figure 2.9). There are four fundamental methods of achieving human
comfort by cooling. They may be applied separately or in combination.

1. Lower the ambient air temperature.

2. Reduce the humidity if it is high.

3. Increase the humidity by evaporation when it is initially low.

4. Use air movement.

The Human Comfort Chart has two comfort zones. The smaller is the
range of effective temperatures over which 50 percent of the people tested
felt comfortable. It is divided into summer and winter zones by minimum
requirements for comfort without mechanical assistance. The larger zone
is the ventilation comfort zone, the range in which air movement is effective in creating human comfort. Within it, minimum air velocities establish the comfort threshold.

Air movement with respect to air quality, energy, and comfort may be
either an ally or an adversary (Table 2.2). The architectural designer
must, therefore, comprehend its advantages and disadvantages.

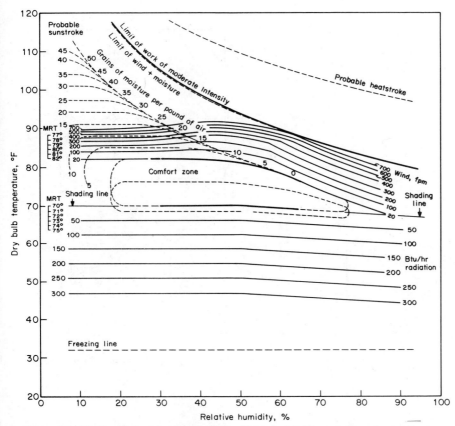

Figure 2.7 Bioclimatic chart developed by Victor Olgyay for inhabitants in the moderate-climate areas of the United States. [7]

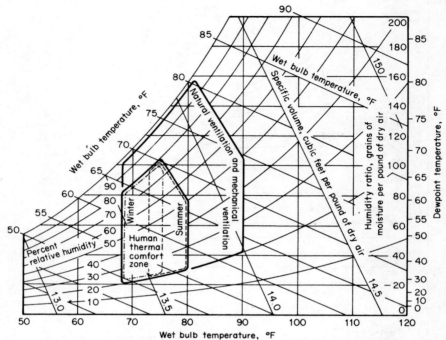

Figure 2.8 Chart of two types of comfort zones: the human thermal comfort zone and the natural ventilation and/or mechanical ventilation zone. [9]

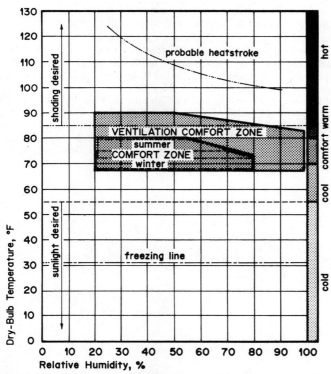

Figure 2.9 The Human Comfort Chart contains two zones: a summer-winter comfort zone and a ventilation comfort zone.

TABLE 2.2   Characteristics of Air Movement

Advantages	Disadvantages
Cools	Cools
Warms	Warms
Dilutes smoke and odors	Spreads smoke and odors
Evaporates moisture and dries surfaces	Aids water to penetrate surfaces of buildings
Prevents cold air from sinking at night	Assists in snow accumulation
Is a source of power	Requires tall buildings to be strengthened
Ventilates	

REFERENCES

1. B. Givoni, *Man, Climate, and Architecture,* Applied Science, Ltd., 1976.
2. William A. R. Thomson, *A Change of Air,* Scribner's, New York, 1979.
3. Michael Lafavore, "Something's in the Air," *New Shelter,* vol. 3, no. 5, May/June 1982.
4. Harvy Sachs, "Clearing the Air," *New Shelter,* vol. 5, no. 7, September 1984.
5. Edward A. Arens and Philip B. Williams, "The Effect of Wind on Energy Consumption in Buildings," *Energy and Buildings,* vol. 1, no. 1, May 1977.
6. American Society of Heating, Refrigerating, and Air-Conditioning Engineers, *ASHRAE Handbook—1985 Fundamentals,* ASHRAE, New York, 1985.
7. Victor Olgyay, *Design with Climate,* Princeton University Press, Princeton, N.J., 1963.
8. Frederic S. Langa, "Many Ways to Cut It," *New Shelter,* vol. 3, no. 6, July/August 1982.
9. Donald Watson, *Energy Conservation through Building Design,* McGraw-Hill, New York, 1979.

3

Methods of Studying Air Movement

To comprehend the movement of air about buildings, one must acquire an appreciation for the techniques of studying air movement. The study involves fluid dynamics and requires a certain amount of diligence. To understand the methods used, it is natural to turn to aerodynamics; unfortunately, few studies prove to be useful. However, since mathematics and wind tunnel models are valuable study tools in aerodynamics, the same methods can be applied to the study of air movement control. The recent introduction of computers combines mathematics and visual graphics. Those three methods—mathematics, computers, and wind tunnel models—are the ones used.

Mathematics

Mathematics is the oldest method of studying air movement; Sir Isaac Newton used it in the 1600s. Five separate equations enable the architectural designer to examine ventilation rates.

Ventilation by pressure forces is influenced by the actual air movement velocity, prevailing direction, variations in intensity and patterns, and local obstructions (Figure 3.1). The equation for pressure force ventilation accounts for these influences with the exception of local obstructions. It determines the quantity of air that passes through inlet openings due to wind forces.

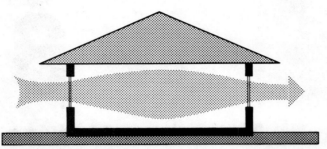

Figure 3.1 The forces of pressure differentials create air movement through a given space.

$$Q = \text{cf } C_v A v \text{ [1]} \tag{3.1}$$

where Q = airflow, cu ft per min
A = free area of inlet opening, sq ft
v = wind speed, mph
C_v = opening effectiveness
cf = conversion factor, 88.0

C_v = 0.50 to 0.60 for wind perpendicular to the inlet opening or 0.25 to 0.35 for wind skewed to the inlet opening; cf converts v to feet per minute.

Equation (3.1) gives the amount of air that flows through a given space, based on optimum conditions. It assumes that the outlet and inlet openings are equal in size so that a constant airflow is maintained. Should the outlet opening be greater in size than the inlet opening, the air movement will be greater; similarly, if the outlet opening is smaller than the inlet opening, the air movement will be smaller (Table 3.1).

TABLE 3.1 Constant of Proportionality Adjustments Due to Pressure

Ratio of inlet area to outlet area	Multiplier of C_v	Ratio of inlet area to outlet area	Multiplier of C_v
1 : 1	1.00	1 : 5	1.40
1 : 2	1.27	2 : 1	0.63
1 : 3	1.35	4 : 1	0.35
1 : 4	1.38	4 : 3	0.86

Determine the ratio of the inlet opening to the outlet opening area. Locate the ratio on the table, and read to the right to find the multiplier of C_v. For example, with a free area inlet opening of 45 sq ft and a free area outlet opening of 135 sq ft, the ratio is 45 : 135 = 1 : 3. Reading to the right, the multiplier of C_v will be 1.35, and the final equation is $Q = (\text{cf})1.35 C_v A v$.

SOURCE: Adapted from Victor Olgyay, *Design with Climate,* Princeton University Press, Princeton, N.J., 1963, p. 104.

Ventilation through buoyancy can be hindered only by the building's internal resistance to air movement if the thermal conditions for buoyancy to occur are correct. For the forces of buoyancy to act, there has to be a significant temperature differential between the interior space and outdoors (Figure 3.2). The following equation is used to calculate the quantity of air through the inlet openings due to buoyancy.

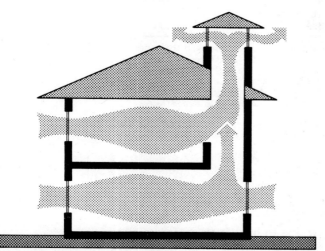

Figure 3.2   Ventilation within a given space may be created through temperature differentials (buoyancy).

$$Q = \text{cf } A[h(T_i - T_o)/T_i]^{1/2} \text{ [1]} \tag{3.2}$$

where Q = airflow, cu ft per min

A = free area of inlet opening, sq ft

h = half the height between inlet and outlet, ft

T_i = average indoor air temperature at height h, °F + 460

T_o = outside air temperature, °F + 460

cf = conversion factor, 313

The conversion factor will be 264 if conditions are not favorable (Table 3.2); T_i represents a cf value of 65 percent for opening effectiveness.

It is assumed that the inlet and outlet openings are equal in size. Should the outlet opening be larger than the inlet opening, air movement through the space will be greater, and vice versa.

Usually, ventilation is due to both pressure and buoyancy forces causing air movement through a given space greater than either force could if it were acting alone. However, the new air movement is not the sum of the two forces; it is the square root of the sum. [1]

TABLE 3.2 Constant of Proportionality Adjustments Due to Buoyancy

Ratio of inlet area to outlet area	Multiplier of cf	Ratio of inlet area to outlet area	Multiplier of cf
1 : 1	313	1 : 5	432
1 : 2	394	2 : 1	197
1 : 3	417	4 : 1	107
1 : 4	429	4 : 3	264

Determine the ratio of the inlet opening area and the outlet opening area. Locate the ratio on the table, and read to the right to find the multiplier of cf. For example, with a free area inlet opening of 30 sq ft and a free area outlet opening of 150 sq ft, the ratio is 30 : 150 = 1 : 5. Reading to the right, the multiplier of cf will be 432 and the final equation is $Q = (432)A[h(T_i - T_o)/T_i]^{1/2}$.

SOURCE: Adapted from Victor Olgyay, *Design With Climate*, Princeton University Press, Princeton, New Jersey, 1963, p. 112.

$$Q = \sqrt{Q_p + Q_b} \tag{3.3}$$

where Q = total airflow, cu ft per min
Q_p = airflow due to pressure, cu ft per min
Q_b = airflow due to buoyancy, cu ft per min

When any variations in the airflow are caused by changes in such factors as opening size, opening effectiveness, or drive of the force, total airflow performance is increased or decreased.

A mathematical formula to calculate wind velocities is based on the averages of occurrences. Wind velocities and the averages of wind prevalence are computed by the averages of the peak winter months and the peak summer months. The wind score equation determines the direction a residence should face for maximum exposure to air movement when prevailing winter and summer winds occur.

$$W_{st} = - (P_1 V_1 C_1) [+ V_2 - (V_2 C_v \div P_2)] \tag{4}$$

$$\{P_2 V_2 \div P_2 \div [V_2 - (V_2 C_v \div P_2)]\} C_2^{[2]}$$

where W_{st} = resultant wind score, deg. The wind directions are as follows:
0° and 360° are north; 90° is east; 180° is south; and 270° is west
P_1 = average winter wind prevalence, percent
P_2 = average summer wind prevalence, percent
V_1 = average winter wind velocity, percent
V_2 = average summer wind velocity, percent
C_1 = thermal winter coefficient

C_2 = thermal summer coefficient
C_v = variable velocity coefficient

When the average thermal characteristic of winter is mild, C_1 = 1.00; when it is cool, C_1 = 1.25; and when it is cold, C_1 = 1.50. When the average thermal characteristic of summer is mild, C_2 = 1.00; when it is warm, C_2 = 1.25; and when it is hot, C_2 = 1.50. C_v is the difference between the average summer and winter wind velocity.

The equation is based on the assumption that summer winds are desirable and winter winds are undesirable. If winter winds as well as summer winds are desirable, the negative sign preceding the average winter wind prevalence symbol P_1 should be changed to a positive sign.

Heat loss from a residential structure is affected by several factors the main one of which is temperature differential between opposite sides of a wall. Air movement also affects heat loss. Equations (3.5) and (3.6) illustrate the influence of the two factors. First, Equation (3.5) is used to determine the total heat transfer through a structural system such as a wall or roof:

$$Q = UA(t_1 - t_2)^{[3]} \tag{3.5}$$

where Q = rate of heat flow, Btu per hour
U = overall coefficient of transmission, Btu per hr per sq ft per °F
A = area of construction having the coefficient U, sq ft
t_1 = temperature of warmer side, °F
t_2 = temperature of cooler side, °F

The coefficient of transmission U varies with different air velocities that are in contact with the building's skin.

The formula for the coefficient of transmission is determined by

$$U = 1 \div \{R_i + R_w + [R_r R_c \div (R_r + R_c)]\}^{[4]} \tag{3.6}$$

where U = overall coefficient of transmission, Btu per hr per sq ft per °F
R_i = resistance of interior surface
R_w = resistance of wall
R_r = resistance of exterior surface radiation
R_c = resistance of exterior surface convection

Air movement, wind in particular, affects the resistance of the exterior surface convection R_c in the derivative of the coefficient of transmission U. As the air velocity increases along the surface of the building, the still air film on the surface is disturbed through turbulent mixing of the air, and the result is increased heat loss from the building. The rate of heat loss increases directly as the rate of air movement over the building surface increases and the roughness of the surface decreases. [3] Therefore, as the air movement about the surface of the building increases, the resistance of the exterior surface decreases with regard to convection heat loss R_c and the overall coefficient of transmission U increases in value. Heat loss due to convection is more significant with regard to the insulative value of the building material, such as single-pane glass compared to insulated glass. Although masonry, because of its rough texture, has half the resistance of single-pane glass to heat loss through convection, the masonry overall coefficient of transmission U is 4 times lower than the single-pane glass because of its internal resistance (Figure 3.3). The resistance of external surface radiation R_r on the other hand, is not affected by air movement. It is influenced by the emissivity and temperature of the building surface.

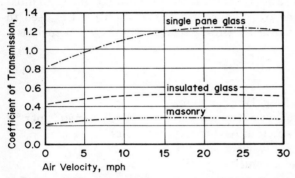

Figure 3.3 The resistance of exterior surface convection R_c is reduced by greater air movement which results in an increased value of the overall coefficient of transmission U. Consequently, the structure's heat loss is larger. [4]

High-emissivity surfaces, such as dull, dark roofs, emit or receive heat readily, whereas low-emissivity surfaces, such as bright, shiny roofs, reflect heat.[3] As the temperature differential between two points on opposite sides of a building such as a wall increases, the radiation heat loss increases rapidly. [3] The resistances of the wall R_w and the interior space surfaces R_i, like radiation, are unaffected by air movement. Both the total heat transfer Q and the coefficient of transmission U are impacted on by air movement through the exterior surface convection characteristics of the building envelope.

The reduction of heat gain through air movement is not found by mathematics, since it deals with the worst thermal condition—a still, hot day. However, thermal barriers and forced air movement may reduce the amount of heat within a structure.

Generally speaking, mathematics may be an easy method of determining specific values for certain functions with regard to air movement. It is particularly useful in evaluating air movement velocities through given inlet openings.

Computers

The quality and usefulness of computers have increased rapidly with advances in technology and the development of software systems for analyzing air movement with respect to architectural structures. United Kingdom universities have developed two such systems. The systems are still being expanded and refined with the integration of data on radiation intensity and sunshine availability.

The control volume and vortex systems were developed as two-dimensional simulations of actual air movement. Several test runs are performed on a particular building or group of buildings to establish airflow direction and magnitude. These time-lapse results are averaged into a final solution. Further refined knowledge of distributions and velocities of pressures with respect to air movement is required for advanced analysis of architectural structures. Presently computer simulation appears to offer no such capabilities.

Wind Tunnels

Wind tunnels and models were extensively investigated by the Texas Engineering Experiment Station during the 1950s (Figure 3.4). After that, however, little analysis of air movement through and about architectural structures was done because of the popularity of air conditioning. As energy costs soared in the 1970s, there was a ventilation renaissance and wind tunnels were used as major research devices.

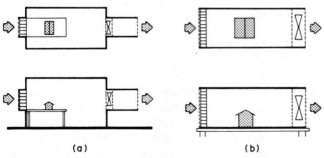

(a)   (b)

Figure 3.4 (a) The Texas Engineering Experiment Station wind tunnel utilizes a table within a room. (b) A typical wind tunnel is, in simple terms, a box with a fan.

Put simply, the wind tunnel is a mechanical means of producing wind that is controlled by a test person to study air movement around a model or models. The tunnel inlet controls the degree of uniformity of air movement through the wind tunnel. Egg crate louvers, insect screens, and cheesecloth are often used at the inlet. A variable-speed fan is located at the outlet. In addition, baffles, insect screens, and/or cheesecloth may be used at the outlet to ensure uniform airflow. Careful calibration of the wind tunnel is essential to accurate analysis.

Means of tracking the airflow include kerosene smoke, titanium tetra-chloride smoke, methane-filled toy balloons, soap bubbles, and whirling trolleys (Figure 3.5). The first two are more successful in making the airflow pattern visible. The air velocity through the tunnel is set between

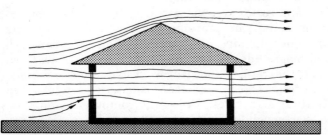

Figure 3.5 Experimentally produced lines that illustrate air movement velocities and pressure zones. When the lines are close together, the air velocity is high and the pressure is low; when the lines are spread apart, the air velocity is low and the pressure is high.

stagnation and destruction of the test model, usually from 60 to 600 fpm.[5] The velocities in and around the test models can be measured by hot-wire anemometer probes. Multiple-velocity measurements can be made with several probes, one probe for each location. In a smoke chamber, which is a scaled-down version of the wind tunnel, probes are difficult to use. Visual observations and/or photographs of the smoke patterns are relied on to identify the area being ventilated and the airflow patterns. Airflow velocities in a smoke chamber are estimated from the relations of airstreams. Once the tests are completed, the data gained are analyzed.

Models are usually constructed of cardboard, acetate, paper, Plexiglass, and/or balsa wood with pins or glue. They are made to a scale ranging from ⅟₁₆ to 1 in. equals 1 ft. Any scale will work as long as careful attention is paid to the accuracy of dimensional and operational characteristics. Most of the time consumed in wind tunnel studies is in model construction. Actual testing of the models involves little time, and modifications to existing models require only a small amount of time. In addition, the effects of variations in wind direction are studied by turning the model to various angles.

Utilization of wind tunnels provides data at several levels. Not only two- but also three-dimensional impressions of airflow can be obtained. Smoke chambers, on the other hand, are almost limited to two-dimensional studies, which are easily converted into three-dimensional information. In addition, wind tunnel studies are sensitive to minor details in buildings. The greatest advantage of wind tunnels is that the model or models respond quickly to multiple variations and numerous factors and can be easily modified.

Comparisons of Study Methods

With three different methods of studying air movement available, an evaluation of the advantages and disadvantages of each is essential. Five separate aspects are compared: accuracy, economics, flexibility, dimensions, and sensitivity. The major concern is to obtain accurate results economically.

No results are valuable unless there is some degree of accuracy in the final outcome. Equations provide a high degree of accuracy but have their limitations. First, they determine air changes which occur in a given space and not the airflow. If air movement cannot be felt regardless of the number of air changes that occur, no thermal relief is gained. Second, they do not account for the patterns of air movement. For example, if a given space has optimum conditions for natural ventilation but air movement occurs only across the ceiling, the occupants will not feel any relief. Mathematical analysis, that is, cannot be used to precisely determine the effectiveness of ventilation. Elmer G. Smith, a research physicist, determined that, with wind tunnel studies, "irregular variations are of the order of 10 percent or less" and probably have "no important change in the pattern" of the airflow.[5] Computer programs are still in the experimental stages, and their level of accuracy has not yet been determined. Air movement simulation with computers may replace mathematics and wind tunnels, but only in the future if ever. All in all, wind tunnel studies appear to be superior to the other methods at the present time.

Economics plays an important role in the endeavor to study air movement. Costs are measured by actual expenses and labor-hours. Mathematics requires the least amount of actual expenses; it may be utilized by a person with only a pencil and paper. Mathematics is the most economical method of studying air movement. The cost of wind tunnel studies involves construction of the wind tunnel, calibration of the wind tunnel, purchase of possible elaborate instrumentation, and construction of the test models. Labor-hours spent in setting up a wind tunnel usually equal the actual expenses of the tunnel and equipment (Figure 3.6).

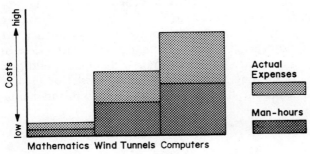

Figure 3.6 The economics of study methods, consisting of actual expenses for material and equipment and man-hours utilized.

The expenses and labor-hour costs are greatly lowered for test runs. A typical run involves only a few minutes of time and the electricity to run a simple fan. Computer costs are much greater; they include purchasing the computer and the considerable amount of electricity required for a single test run. Numerous labor-hours are consumed in developing computer programs, programming the computer, and analyzing the computer printout. In short, mathematics is the most economical method of studying air movement, followed by wind tunnels. However, the type of results obtained must be compared to the expenses sustained in determining the cost-benefit of the study method. In that light, wind tunnel studies acquire the largest quantity of information for the money spent.

Flexibility is essential to the study of air movement, since alterations during the study process are inevitable. In that respect, mathematics is limited. It deals with specific rules, a certain number of factors, and only a few variations. Wind tunnel models are inexpensive and easy to modify (Figure 3.7). Multiple variations are possible: numerous factors can be

Figure 3.7 The flexibility of wind tunnel studies allows for numerous variations and alterations.

introduced to a wind tunnel study by adding landscaping, fences, hills, and buildings and changing the orientation of the model.

Computers, like mathematics, have limited flexibility; they use equations as a starting point and a matrix system that restricts their ability to change except at large scale. Although the building orientation may be altered, the mathematical origin limits the computer to a specific number of factors and few variations. Flexibility is inherent in wind tunnel studies.

Air movement is usually investigated in two or three dimensions, which mathematical equations cannot do because the solutions are purely numerical. Numbers deal only with values or quantities, not with spaces or volumes. Wind tunnel results are three-dimensional because the models are exact miniature replicas of full-size structures. Computer responses are only two-dimensional. Although the extension of computer technology into three-dimensions is "relatively straightforward," the method of conversion is quite crude and is in need of refinement. Wind tunnel studies represent the actual situation as closely as possible.

Details are the fine tuning of architecture and air movement control. Unfortunately, mathematics does not respond to details that may occur in buildings and the division of spaces in buildings is not detected by mathematical equations. The Texas Engineering Experiment Station's research physicist, Elmer G. Smith, discovered that "small changes in a structure may cause large changes in the airflow pattern." [5] This fact was noticed in a series of wind tunnel tests (Figure 3.8). As a result, replica

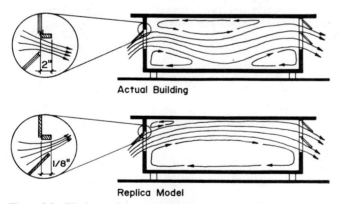

Actual Building

Replica Model

Figure 3.8 Wind tunnel tests are sensitive to minor details. The actual building and the replica model were intended to be the same for a series of experiments performed at the Texas Engineering Experiment Station. The inner edge of the window projected slightly farther inside the wall of the model than it did in the actual building and created an unusual airflow pattern. When the window position was slightly moved outward from the wall, less than $\frac{1}{8}$ in, the same pattern was obtained in the model as in the actual building. [5]

models and simplistic models were found to be very accurate in creating the exact air movement patterns in actual buildings. Wind tunnel study method proves to be sensitive to minute details. In contrast, computers are based on a large matrix cell system and a typical cell could easily contain a whole house.

Summary

Although wind tunnels are less economical than mathematics, they are more cost-beneficial. They explore a large number of factors and variations, which neither mathematics nor computers can do. With their sensitivity to details, they are superior to the other methods. Flexibility is another important quality of the wind tunnel method, as is the models being created as exact replicas of actual buildings. Even the air movement patterns produced around a model are identical to those around the full-size structures.

REFERENCES

1. American Society of Heating, Refrigerating, and Air-Conditioning Engineers, Inc., *ASHRAE Handbook—1985 Fundamentals,* ASHRAE, New York, 1985.
2. Victor Olgyay, *Design with Climate,* Princeton University Press, Princeton, N.J., 1963.
3. Bertram Y. Kinzey, Jr. and Howard M. Sharp, *Environmental Technologies in Architecture,* Prentice-Hall, Englewood Cliffs, N.J., 1963.
4. Edward A. Arens and Philip B. Williams, "The Effect of Wind on Energy Consumption in Buildings," *Energy and Buildings,* vol. 1, no. 1, May 1977.
5. Elmer G. Smith, *The Feasibility of Using Models for Predetermining Natural Ventilation,* Research Report No. 26, Texas Engineering Experiment Station, College Station, Tex., 1951.

4

Climatic Data

Climate has always affected architectural structures. In spite of designers' attempts to produce dryness in rainstorms, heat in winter, coolness in summer, acoustic and visual privacy in openness, and mini-environments, climatic conditions still influence the interior space of buildings. A structure designed for the Florida climate would not perform well in the Arctic, or vice versa. Even if an architectural capsule that would completely isolate dwellers from the exterior environment were built, it would still respond to the effects of its immediate environment.

Anyway, no one would want to live in it; most persons wish to maintain some kind of contact with nature. Individual performance is dependent on climatic conditions that affect the capacity of the body for physical and mental work, sleep, rest, and overall enjoyment of life. A suitable climate will produce feelings of vigor; an unsuitable climate will stimulate depression; and a lack of climatic conditions "destroys vigor and happiness of, and brings about the atrophy of disuse in, men." [1] Most people do not prefer a single and constant climate; they want climates compatible with their personalities and tastes. Changes in climate stimulate people.

Buildings should buffer their occupants from climatic extremes and accentuate the benefits of the climate. Architectural designers should work with nature. Understanding the general characteristics of different climatic regions as well as those of a particular meso-climate or micro-climate is essential to developing a satisfactory building design.

Climatic Regions

Each climatic region has its own set of weather-related problems (Figure 4.1). Human discomfort can range from smothering desert heat to sticky humidity. However, comfort can be obtained in each region if the

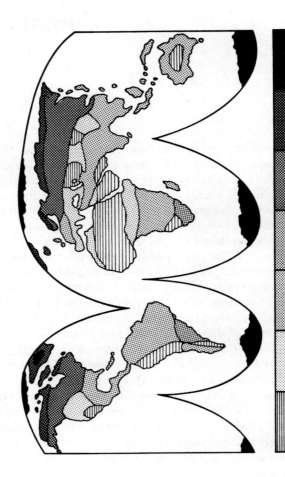

Desert Hot & Arid Hot & Humid Composite Cold & Humid Cold & Arid Ice Cap

Figure 4.1 Major climatic regions create an organized pattern across the world.

designer understands the specific climate and develops a working knowledge of that climate. The seven major climatic regions are not precisely defined; they flow from one area to another. The overlap areas are called transition zones; the climate in those zones is a mixture of the adjoining climatic regions. Although the seven major regions may shift slightly from year to year or month to month, they do remain relatively stable and confined.

The desert climatic region is characterized by very hot summers, very cold winters, and an extremely large daily fluctuation of temperatures. Low humidity is evidenced by the rare occurrence of precipitation. Air movement is minor except for occasional windstorms of severe intensity.

The hot and arid climatic region, a milder version of the desert climate, has hot summers, cold winters, and high daytime and low nighttime temperatures. This climatic region is marked by low humidity, small amounts of precipitation, and little air movement.

The hot and humid climatic region has warm-to-hot summers, cool winters, a relatively small daily fluctuation of temperature, and little seasonal variations. Precipitation is high in all seasons and creates a high level of humidity. Light air movement, often interrupted by long calm periods, is typical.

The composite climatic region has two distinct seasons. The warm-to-hot summers are low in humidity, and precipitation is rare. The cool winters are a definite opposite to the summers: the humidity is high, and precipitation occurs often. This region, with its relatively small daily temperature fluctuation, has frequent and moderate air movement.

The cold and humid climatic region is marked by warm summers, cool springs, long cold winters, a small daily temperature fluctuation, and high humidity. Some precipitation occurs year-round, and summer breezes have relatively reliable directional characteristics.

The cold and arid climatic region is distinguished by its short cool-to-warm summers and long cold winters. Small daily temperature fluctuation, low humidity, and slight precipitation describe this region thoroughly. Air movement occurs occasionally.

The ice cap climatic region has cold summers and extremely cold winters. The barely detectable daily temperature fluctuation and low humidity further enhance the sensation of cold in the region. In addition, air movement is quite a common occurrence.

Comparisons of Study Methods

Within each climatic region are several meso-climates with their own characteristics. This level is the first at which air movement can be analyzed, although the most effective study is at the micro-climatic level.

TABLE 4.1 Sample Climatic Data

Time	Factor	\multicolumn{15}{c}{Date}

Time	Factor	1	2	3	4	5	6	7	8	9	10	11	12	13	14	15
4 A.M.	Temperature	28	36	35	37	44	40	49	41	39	57	55	39	41	42	49
4 A.M.	Humidity	69	73	92	86	77	89	100	70	86	93	87	89	96	100	96
10 A.M.	Temperature	40	48	49	47	51	52	53	50	57	64	50	38	41	43	49
10 A.M.	Humidity	47	48	50	66	77	57	69	50	69	97	96	89	93	100	96
4 P.M.	Temperature	55	62	63	62	65	67	65	64	66	71	44	45	47	49	51
4 P.M.	Humidity	37	52	40	34	43	23	27	29	65	79	86	93	96	86	93
10 P.M.	Temperature	37	46	45	40	44	53	45	43	59	61	42	42	44	49	49
10 P.M.	Humidity	76	93	71	86	83	83	66	86	81	84	93	93	96	96	93

Temperature is taken in degree Fahrenheit and humidity in percentage; NA indicates data not available.
SOURCE: Courtesy of National Oceanic and Atmospheric Administration and National Climatic Data Center, *Local Climatological Data, Monthly Summary*, U.S. Department of Commerce, National Climatic Data Center, Asheville, N. C., 1984.)

Analysis at the meso-climatic level develops basic principles of air movement control, and techniques for use at the micro-climatic scale are established.

Once the information is collected, it must be put into a format that designers can use to obtain comfort for occupants (Table 4.1). The results are based on the combined effects of temperature and humidity, the main factors in comfort or discomfort. Air movement and sky coverage are

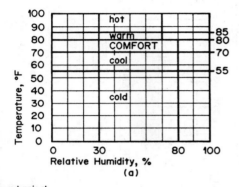

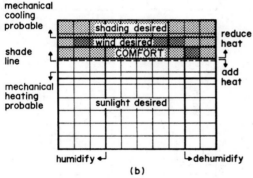

hot/ dry	hot/ normal	hot/ humid
warm/ dry	warm/ normal	warm/ humid
dry	COMFORT	humid
cool/ dry	cool/ normal	cool/ humid
cold/ dry	cold/ normal	cold/ humid

(c)

Figure 4.2 The temperature and humidity chart (a) is based on thermal comfort. Superimposed on (a) are various climatic factors as seen in the climate impact chart (b). The temperature and humidity chart is converted into a 15-cell matrix system (c) allowing easier utilization in architectural decisions. The relative humidity with respect to comfort changes from 80 to 70 percent above the 80°F temperature line because of the increased effect of humidity in the higher temperature zones. *(Adapted from Ref. 2)*

Figure 4.3 The climate-comfort analysis chart for a specific area illustrates the average monthly temperature and humidity effects on comfort. The values shown are percentages of time.

Data tables for Figure 4.3 (columns: Dry | Norm | Humid; rows: Hot, Warm, Comfort, Cool, Cold):

JANUARY
	Dry	Norm	Humid
Hot			1
Warm		3	3
Comfort		9	20
Cool		23	38
Cold	3		

FEBRUARY
	Dry	Norm	Humid
Hot			
Warm		10	1
Comfort		19	32
Cool		18	16
Cold	3		

MARCH
	Dry	Norm	Humid
Hot		2	
Warm		16	5
Comfort		30	23
Cool		8	14
Cold	2		

APRIL
	Dry	Norm	Humid
Hot		1	
Warm		20	5
Comfort		26	27
Cool			11
Cold			

MAY
	Dry	Norm	Humid
Hot		7	
Warm		16	1
Comfort		20	2
Cool		8	17
Cold			

JUNE
	Dry	Norm	Humid
Hot		15	
Warm		25	36
Comfort		13	9
Cool		1	1
Cold			

JULY
	Dry	Norm	Humid
Hot		6	1
Warm		25	4
Comfort		5	53
Cool			6
Cold			

AUGUST
	Dry	Norm	Humid
Hot		32	2
Warm		11	3
Comfort		4	44
Cool			4
Cold			

SEPTEMBER
	Dry	Norm	Humid
Hot		12	2
Warm		17	3
Comfort		10	42
Cool		2	16
Cold			

OCTOBER
	Dry	Norm	Humid
Hot			
Warm		16	4
Comfort		22	26
Cool		4	27
Cold			1

NOVEMBER
	Dry	Norm	Humid
Hot			
Warm		2	
Comfort		12	7
Cool		21	25
Cold		8	25

DECEMBER
	Dry	Norm	Humid
Hot			
Warm		18	5
Comfort		8	52
Cool		5	12
Cold			

Figure 4.4 Increased sunshine or reduced sky coverage raises the average monthly temperature. The values shown are percentages of time. Shading is an important factor in creating human comfort as seen by comparing this chart with Figure 4.3.

Data tables for Figure 4.4 (columns: Dry | Norm | Humid; rows: Hot, Warm, Comfort, Cool, Cold):

JANUARY
	Dry	Norm	Humid
Hot			1
Warm		3	3
Comfort		9	20
Cool		23	38
Cold	3		

FEBRUARY
	Dry	Norm	Humid
Hot			
Warm		10	1
Comfort		19	32
Cool		18	16
Cold	3		

MARCH
	Dry	Norm	Humid
Hot		2	
Warm		16	5
Comfort		30	23
Cool		8	14
Cold	2		

APRIL
	Dry	Norm	Humid
Hot		11	
Warm		20	5
Comfort		26	27
Cool			11
Cold			

MAY
	Dry	Norm	Humid
Hot		23	
Warm		2	20
Comfort		8	26
Cool		17	3
Cold			

JUNE
	Dry	Norm	Humid
Hot		40	1
Warm		13	36
Comfort		1	9
Cool			1
Cold			

JULY
	Dry	Norm	Humid
Hot		31	5
Warm		5	53
Comfort			6
Cool			
Cold			

AUGUST
	Dry	Norm	Humid
Hot		43	5
Warm		4	44
Comfort			4
Cool			
Cold			

SEPTEMBER
	Dry	Norm	Humid
Hot		29	1
Warm		10	42
Comfort		2	16
Cool			
Cold			

OCTOBER
	Dry	Norm	Humid
Hot			
Warm		20	
Comfort		22	26
Cool		4	27
Cold			1

NOVEMBER
	Dry	Norm	Humid
Hot			
Warm		2	
Comfort		12	7
Cool		21	25
Cold		8	25

DECEMBER
	Dry	Norm	Humid
Hot			
Warm		18	5
Comfort		8	52
Cool		5	12
Cold			

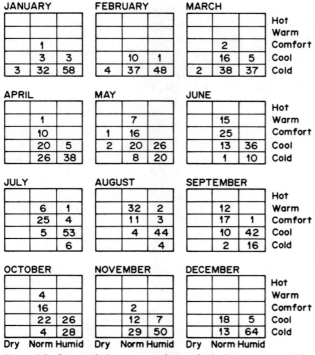

Figure 4.5 Increased air movement lessens both the average monthly temperature and the humidity. The values shown are percentages of time. Note that the hot and warm categories lack high percentage values as compared with Figures 4.3 and 4.4.

utilized as secondary factors by either reducing or increasing the effective temperature. In addition, devices that create shade decrease the ambient air temperature. Precipitation raises the humidity. The effects of the secondary factors are seen in the analysis (Figures 4.2 to 4.5). The analyzed data of a specific climate, meso-climate or microclimate, are compared to the information in the *Human Comfort Chart* (Figure 2.9). The result is a climate-comfort analysis chart (Figure 4.3) which can be used directly by the designer.

A matrix of the temperature and humidity chart allows the architectural designer to determine the atmospheric quality of a given region. Data presented in the matrix are in percentages of time. From the chart, the designer can easily find which environmental conditions must be overcome to achieve human comfort.

REFERENCES

1. William A. R. Thomson, *A Change of Air*, Scribner's, New York, 1979.
2. National Oceanic and Atmospheric Administration and National Climatic Data Center, *Local Climatological Data Monthly Summary*, U.S. Department of Commerce National Climatic Data Center, Asheville, N.C., 1984.
3. The Bureau of Research, *Houses and Climate: An Energy Perspective for Florida Builders*, The Governor's Energy Office, Tallahassee, Fla., 1979.

5

Meso-climatic Air Movement

Architectural designers who understand the major characteristics of air movement at the meso-climatic scale can incorporate environmentally responsive flexibility in their designs. The principles of fluid dynamics are best understood by getting into the techniques of air movement control at the micro-climatic level. Air movement is a three-dimensional phenomenon, and its complexity is best understood by viewing it that way.

Internally, air movement is driven by buoyancy and pressure differentials and modified by inertia and friction. Externally, landforms, vegetation, and buildings influence air movement by altering the velocity and pattern of the airflow. Air movement changes affect the quality of the environment at both the meso-climatic and micro-climatic levels, as well as the energy consumption of structures. The environmental factors of a given climate affect the movement of air, and, in turn, air movement affects the climate (Table 5.1).

General Principles

As air circulates in its three-dimensional sphere, it follows specific laws of nature. These regulators assist and drive the movement of air in a relatively orderly and predictable fashion. Air movement is a cause-and-effect phenomenon; and if the designer understands the effect of air movement velocities and patterns, he or she may also locate the causes.

The flow patterns of air movement fall into three categories (Figure 5.1). In *laminar air movement*, which occurs quite frequently, airstreams flow on top of or beside one another in a relatively parallel and predictable

TABLE 5.1 Characteristics of Various Wind Speeds

Feet per minute	Miles per hour	General description	Specifications
Less than 88	Less than 1	Calm	Smoke rises vertically.
88–264	1–3	Light air	Wind direction shown by smoke drift but not by vanes.
352–616	4–7	Slight breeze	Wind felt on face; leaves rustle; ordinary vane moved by wind.
704–968	8–11	Gentle breeze	Leaves and twigs in constant motion; wind extends light flag.
1056–1408	12–16	Moderate breeze	Dust and loose paper moved; small branches are moved.
1496–1936	17–22	Fresh breeze	Small trees with leaves begin to sway.
2024–2376	23–27	Strong breeze	Large branches in motion; whistling in telephone wires.
2464–2992	28–34	Moderate gale	Whole trees in motion.
3080–3608	35–41	Fresh gale	Twigs broken off trees; progress generally impeded.
3696–4224	42–48	Strong gale	Slight structural damage occurs; chimney pots removed.
4312–4928	49–56	Whole gale	Trees uprooted; considerable structural damage.
5016–5896	57–67	Storm	Very rarely experienced; widespread damage.
Above 5984	Above 68	Hurricane	Extremely rare; extensive damage common.

Air movement speeds are measured at 20 ft above the ground.

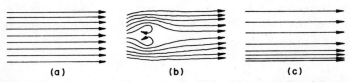

Figure 5.1   Air movement patterns are classified as (a) laminar, (b) turbulent, and (c) separated.

pattern because of low internal turbulence. As external elements increase the internal turbulence, the pattern of the airstreams becomes random and unpredictable. The main path, however, occurs in one major direction and creates *turbulent air movement*. Then friction may decrease the velocity of certain airstreams while maintaining a parallel pattern without internal turbulence. The result is *separated air movement*. [1] Air movement may change from one category to another over a period of time or distance (Figure 5.2). Laminar air movement, for example, may become turbulent air movement if the topographical roughness becomes greater. Buildings, as well as landforms, create turbulent and separated air movement as the airstreams pass them.

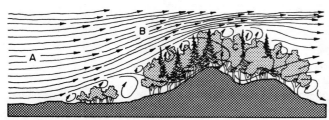

Figure 5.2   Air movement patterns change from one classification to another depending on internal and external factors. Here air movement varies from (*a*) laminar to (*b*) separated to (*c*) turbulent.

The patterns are based on the assumption that the air movement is viewed in either the horizontal or the vertical plane. A particular air movement may fall in one category in the horizontal plane and in another in the vertical plane, or both air movements may be in the same category. In any case, understanding the characteristics of three categories is a valuable tool in predicting air movement patterns.

As air flows from one location to another, inertia, friction, and differential affect the movement. First, moving air has *inertia*: once it is set into motion, it tends to continue in the same direction until it is diverted from its original path (Figure 5.3). Elements that divert the flow are buildings, topography, trees, shrubs, furniture, automobiles, and people. A change in direction will reduce the velocity of the airflow; the greater the initial velocity the greater the reduction.

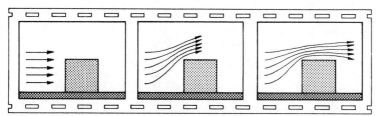

Figure 5.3   Air traveling in a given direction will continue in that direction until it is diverted by an element such as a building or tree.

Second, moving air creates *friction* as it moves along such bodies as land, water, and buildings (Figure 5.4). Friction reduces the velocity of air movement and may even alter airflow patterns. In addition, the friction between the air and the revolving earth produces a gradation of air velocities

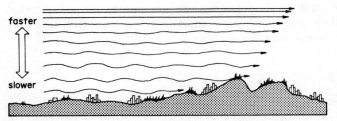

Figure 5.4   Friction reduces the movement of air by creating a drag effect.

from low to high altitudes. The speed of air movement increases with height as the frictional effect of the earth's surface reduces in its intensity (Figure 5.5). Since the earth's surface moves faster at the equator than at the poles, the frictional effect upon the movement of air is less in the higher latitudes. Consequently, air movement velocity increases as the latitudes approach the poles.

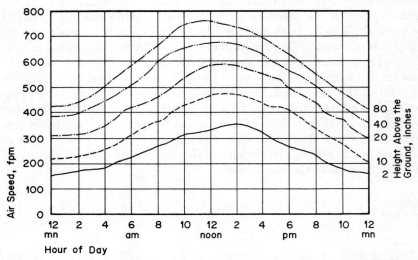

Figure 5.5   Air movement velocity increases with higher altitudes. As the airspeed increases at a given height, the velocity of the air increases proportionally at corresponding heights. *(Adapted from Ref. 2)*

Third, air movement is created by *differentials*. The forces of buoyancy and pressure may work separately or cooperatively in developing differentials. As air passes from positive to negative buoyancy zones and/or from positive to negative pressure zones, air movement is created (Figures 5.6 and 5.7). The gradients of buoyancy consist of variations in air density, and differences in air force action develop the gradients of pressure. Air

up the radiant heat within the slope that has been collected from the sun. On the other hand, a north slope is not heated directly by the sun and air moving on it usually remains cool or cold. If the slope is long enough, it may cause moisture in the air to condense and precipitate on the slope, or it may cause the air to pick up moisture (Figure 5.11). For example, a traveling air mass may pick up moisture from a valley lake as it descends a slope and later that same water vapor may precipitate on a mountainside that the air mass ascends.

Air movement has inherent fluctuations and turbulences even if it occurs over the smoothest surface. Consequently, the rougher the terrain the greater the disturbances within the airflow. That is especially true where the air comes in contact with the ground. However, the main part of the air movement usually continues in the initial direction in spite of minor fluctuations and turbulences.

This information is significant for the architectural designer for two reasons. A structure may be located to receive or shun airflow, or it may be oriented to capture or reflect airflow. To illustrate, a structure may be positioned on a wind-ascending slope to gain and increase the effect of certain desirable winds. Likewise, the same hill may assist in sheltering the building from undesirable winds (Figure 5.12). A valley may be utilized in the same way. Furthermore, the compass position of the hillside can further enhance the quality of air flowing over the surface by creating a warmer or cooler or a drier or damper climatic environment. The ground texture, along with surface friction, may either increase or decrease the effect of a given mass by accentuating or lessening the internal turbulence and initial velocity of the airflow.

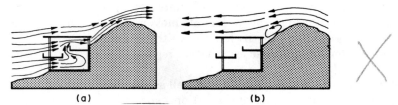

Figure 5.12 Careful positioning and design of a building may invite desirable air movement (a) and shun undesirable airflow (b).

Vegetation

Trees and shrubs do more than enhance a structure's exterior space or increase its real estate value. Air movement can be controlled by vegetation properly selected and placed (Figure 5.13). Filtration, reflection, guidance, and/or obstruction of the airflow can be provided. Vegetation

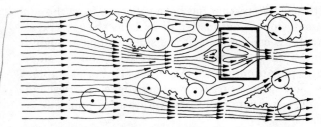

Figure 5.13 Control of air movement may be achieved by proper selection and careful placement of vegetation.

can even reduce or accelerate air movement around buildings and increase or decrease the structure's energy requirement. The effectiveness of the vegetation in controlling air movement depends on the form, density, rigidity, and other vegetation characteristics which vary airflow velocity, pattern, and quality.

Vegetation creates a frictional drag on the airflow, and vegetation with the "optimum density" can reduce the velocity of air movement along the ground "as much as 70 percent." [3] According to research done for the American Society of Landscape Architects Foundation, that optimum density is "about 50 to 60 percent."[1] That is, the leaves, twigs, branches, and trunks of all the vegetation should account for 50 to 60 percent of the volume of the stand (Figure 5.14). The decrease in the flow of air through vegetation is greater when the species are varied (Figure 5.15).

Air movement is deflected over and filtered through vegetation. As moving air approaches a forest stand it is directed upward and over the vegetation for the most part while some is filtered through the vegetation itself. If the stand lacks underbrush, more air will flow under the trees' crowns than through them. The air movement pattern around a forest stand is determined by the length, width, and crown shape of the stand, and the density and species variation regulate the airflow velocity (Figures 5.16 and 5.17).

Vegetation affects the quality of air as well as the air movement itself. As air travels beneath canopies of vegetation, especially trees, it is conditioned. Vegetation can block solar radiation from reaching the area beneath the canopies; consequently, the heat content of the ambient air decreases. The temperature is further reduced by the addition of moisture to the air from the vegetation through a process known as transpiration. The humidity of the air beneath the trees' canopies increases through this process, which even in hot and humid climates, accentuates the cooling sensation.

Vegetation has a wide variety of air movement control potential. The velocity, pattern, and quality of the control may be sensitively altered in minute gradations through proper selection and placement of various species of vegetation. Also, vegetation is not only a potential controller of air movement; it also reduces noise, removes dust particles, absorbs carbon dioxide, and subtracts heat from the air while introducing moisture and oxygen into the air.

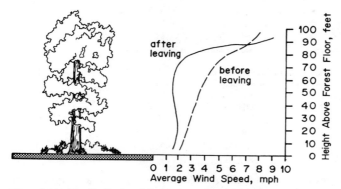

Figure 5.14 The distribution of air movement velocity through an oak grove with beech undergrowth varies with height. The velocity of the airflow is directly proportional to the density of the oak grove. *(Adapted from Ref. 2)*

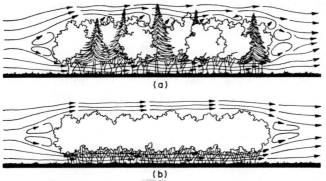

Figure 5.15 The irregularity of a mixed forest stand (*a*) is more effective in reducing the airflow velocity than a more uniform forest stand (*b*).

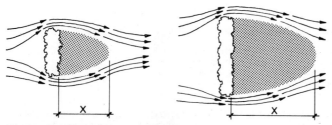

Figure 5.16 The width of a forest stand mainly affects the width of the protected area and not the length, which is determined by the forest stand's height, length, and crown shape.

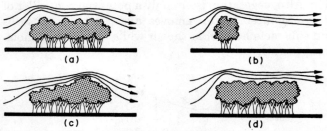

Figure 5.17   The length and shape of a forest stand affect the pattern of air movement over the stand. A long forest stand (a) offers a small protected area behind it, whereas a short forest stand (b) protects a large area. A forest stand with a pitched roof (c) is less effective than a forest stand with a flat roof (d) in redirecting the air movement pattern. [1]

Buildings

Architectural designers are generally aware of the effect of air movement on their structures, but their attention is now being focused on the opposite phenomenon: the effect of buildings on air movement. Structures deflect, obstruct, and guide the movement of air about them as well as reduce and accelerate the airflow velocity. The intensity and magnitude of the influence of buildings on air movement vary with structures' heights, widths, lengths, and forms.

As air movement deflects over and around buildings, distinct air patterns are established. As the air contacts the face of the building on the windward side, it is divided two-thirds up the building's face (Figure 5.18). One-third of the air flows over the top of the building, and two-thirds of the air flows downward. It forms a vortex at the ground and escapes around the building's corners.

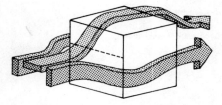

Figure 5.18   A typical air movement pattern around a building form divides the oncoming air two-thirds up the face of the building. One-third of the airflow goes over the top, and two-thirds detours around the sides.

Areas of calm are created on both the windward and leeward sides of the building, where the movement of air is barely detectable (Figure 5.19). The leeward calm area, or jump of the air pattern, is determined by the height, width, and windward face of the building. However, once the width of the building becomes 9 times greater than the height, the width loses its influence on the air pattern and the height becomes the predominant factor. [4]

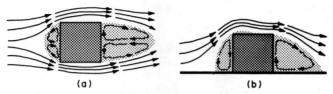

(a)   (b)

Figure 5.19 As the air movement pattern detours around a building, areas of calm, or eddies, are created in both plan (*a*) and elevation (*b*). The eddy on the leeward face of the building is usually larger than the one on the windward face.

Although the building obstructs the airflow and reduces its initial velocity, the changes in airflow pattern increase the air movement velocity at the base and sides of the building (Figure 5.20) as much as "two or even three times the speed in the undisturbed air stream with no building present." [6] However, danger exists only with buildings that are "six stories or more in height and more than twice the height of surrounding buildings." [7]

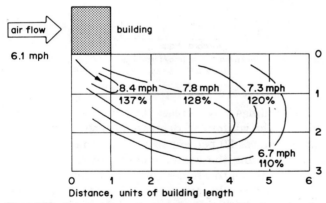

Figure 5.20 In typical corner streams of an average building form, the velocity around the corner increases as much as 137 percent over the initial airflow velocity. [5]

The deflection of the air movement and the reduction of its velocity develop differentials of pressure. Positive pressure is created as air piles up on the windward face of a building or any obstruction (Figure 5.21). The pileup causes the air movement to lose velocity until a new path is located. As the airflow travels around the building in a new pattern, negative pressure areas are formed on the sides and leeward face of the building (Figure 5.22). Both the positive and negative pressure areas lack any significant air movement. The exact pattern of the air movement is determined by the pressure differential as well as by the inertia of the airflow and the form of the building.

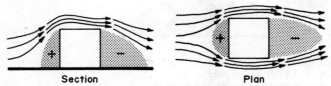

Figure 5.21   Air movement around a building creates positive and negative pressure zones.

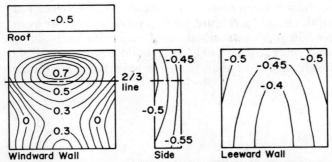

Figure 5.22   Typical pressure distribution for an average building form. The greatest pressure force is at the two-thirds point on the windward wall, where the airflow divides. [5]

Air movement around buildings may increase the energy consumption within the structure during the winter and decrease it in the summer. A house in a windless area requires 30 percent less energy than the same house located in the wind during the winter months. In the summer, a house with a breeze, as compared to a house without a breeze, may not need air-conditioning machinery. Proper air movement control can provide the desired benefits in both circumstances.

Architectural designers who understand the basic principles of air movement control can regulate the flow of air within their structures. Calm and breezy areas are created by utilizing guided airflow patterns while the airflow velocity is either accelerated or reduced. Careful planning establishes air movement control.

REFERENCES

1. Gary O. Robinette and Charles McClennon, *Landscape Planning for Energy Conservation*, Van Nostrand Reinhold, New York, 1983.
2. Rudolf Geiger, *The Climate Near the Ground*, Harvard University Press, Cambridge, Mass., 1959.
3. Jeffrey Ellis Aronin, *Climate and Architecture*, Reinhold, New York, 1953.
4. Louis Allen Harding and Arthur Cutts Willard, *Heating, Ventilating, and Air Conditioning*, Wiley, New York, 1937.
5. A. F. E. Wise, "Ventilation of Buildings: A Review with Emphasis on the Effects of Wind," *Energy Conservation in Heating, Cooling, and Ventilating Buildings*, Vol. 1, Hemisphere Publishing, Hemisphere, Wash., 1978.
6. R. E. Lacy, *Climate and Building in Britain*, Building Research Establishment, London, 1977.
7. Ralph W. Crump, "Games that Buildings Play with Winds," *AIA Journal*, vol. 61, March 1974.

6

Microclimatic Air Movement

The interaction of buildings and the environment is extremely complex, especially in regard to air movement on the micro-climatic scale. As the air flows within the immediate surroundings of buildings, its velocities and the patterns of its movement are reduced, accelerated, obstructed, directed, deflected, and filtered. The utilization of these airflow alterations by the designer may result in climatically comfortable residences.

All residential designs affect the velocities and patterns of airflows, but only climatically designed residences enhance the benefits of such air movement. Air velocity may be influenced to a certain extent, even though it is extremely variable, and airflow patterns, usually predictable, may be shifted and guided within a specific space or building. To achieve significant air movement control, the designer should employ building forms along with surrounding topography, vegetation, fences, and adjacent structures. The complex three-dimensional character of air movement is essentially simple to predict and control once the basic principles and techniques are understood.

Buildings

Architectural structures exist in micro-climates, and the elements of those climates affect the movement of air through the building. The elements may alter the character of the airflow before it reaches the building. However, an investigation of the building with its configurations, orientations, heights, overhangs, roof shapes, and other architectural forms, without the influence of other environmental factors, discloses numerous techniques available to control air movement.

Both the configuration and orientation of a building provide a range of effects on the air movement patterns and velocities (Figure 6.1). The effects alter the structure's interior and exterior environments. In addition, the energy consumption of the building is greatly impacted. Although the catalog of configuration options and orientation variations is large, a few basic air movement control principles may be developed through an investigation of primary building forms.

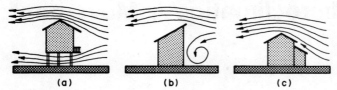

                (a)   (b)   (c)

Figure 6.1   A building's configuration and orientation may (a) guide, (b) obstruct, or (c) deflect air movement velocity and patterns.

The configuration of a structure may increase not only air movement through the structure but also the amount of natural lighting within the building. Architectural forms, such as courtyards, atriums, light wells, wing-walls, dormers, clerestories, porches, balconies, and other creative elements, may enhance both air movement and natural lighting. However, air movement is of prime importance concerning human comfort. Both building configuration and orientation affect the air movement pattern and velocity as they alter the initial airflow.

Five primary building forms are prevalent in current residential designs, and they will simplify the study of air movement (Figure 6.2). Dimensions of the forms are based on units designated by the letter *D*.

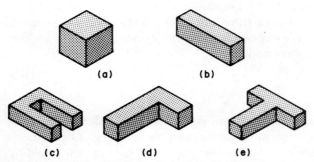

            (a)   (b)

    (c)   (d)   (e)

Figure 6.2   The five primary forms of buildings are (a) square, (b) linear, (c) U-shaped, (d) L-shaped, and (e) T-shaped.

The actual size of *D* is relatively unimportant, since the proportions of the building form determine the size and nature of the air movement pattern (Figure 6.3). For example, a building 25 ft in depth, length, and height will

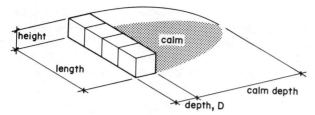

Figure 6.3 Basic dimensions and terminology of the building form. *(Adapted from Ref. 1)*

create a given airflow pattern and an area of calm 50 ft in depth. A building 50 ft in depth, length, and height will create the same pattern as the first building except at a larger scale and an eddy 100 ft in depth. The depth of a calm area, D_c, of a building which has equal depth, length, and height will be twice the building's depth. If $H = L = D$, then $D_c = 2D$.

Calm areas of the five primary building forms reveal some interesting facts about air movement in relation to structure. The eddies in themselves disclose no valuable information except by implication. The size of a calm area, especially depth, is directly related to the potential of air movement through the structure. The structure's degree of obstruction determines its potential effect on internal airflow (Figure 6.4). However, there is a limitation. If a building such as a skyscraper obstructs too much air movement, a gale force airflow

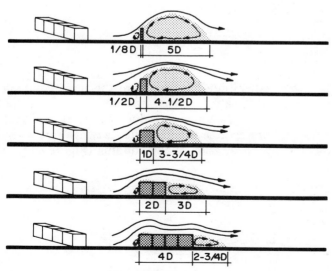

Figure 6.4 A thin mass, such as a wall, provides a greater protected area from air movement than a thicker mass of the same height and width. [1]

may occur inside it (Figure 6.5). The implication of the depth of calm is reinforced by the height of the calm. As the eddy increases in height, the potential for maximum or optimum airflow through the structure increases. Although there are a few exceptions, the size of the calm, including depth and height, is usually an accurate predictor of the air movement potential for the building.

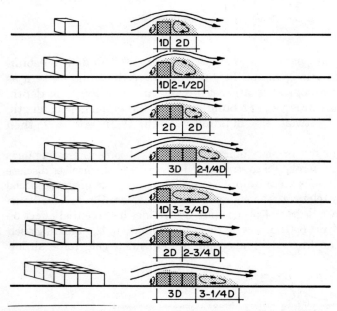

Figure 6.5   The protected area of the calm becomes smaller in proportion to the building as the building becomes larger. [1]

First, square and rectangular building forms are the most common structural shapes. The area of calm decreases proportionately as the building depth is increased. This phenomenon is due to the frictional drag of the roof. The square building creates an eddy which is relatively consistent in size regardless of the orientation of the building (Figure 6.6).

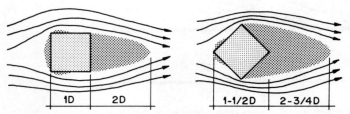

Figure 6.6   The orientation of a square building form does not seem to affect the potential for optimum air movement through the structure.

Second, linear building forms provide a greater opportunity to utilize air movement than a square form provides. The building creates a larger calm area (Figure 6.7). As the linear building is increased in length, the

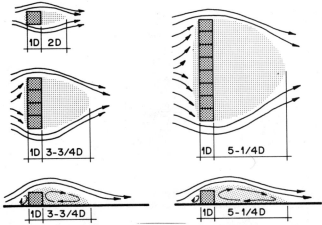

Figure 6.7 As the length of the building is increased, the length and depth of the calm area increase while the height of the calm remains constant. [1]

size of the eddy increases proportionately (Figure 6.8). In addition, a narrow linear building creates a larger eddy than an equally long building that has a greater depth. For example, a structure which is one-third the depth of another building but equal in length has a 40 percent greater protected area. The orientation of a linear building may greatly vary the depth and width of the eddy (Figure 6.9). The greatest area of calm is obtained when the building is slightly skewed, 30° to the airflow. The larger eddy, in comparison to the calm areas of other orientations, indicates that, with proper placement of inlet and outlet openings, an accelerated airflow can be achieved within the structure.

Third, U-shaped building forms react with air movement in a unique manner. Although the area of calm is relatively the same regardless of building orientation, the potential for air movement through the structure varies (Figure 6.10). When the closed end of the U-shape faces the airflow, the building is like a square building; it has a set potential for air movement through it. When the open end of the U-shape faces the air movement, air is collected within the U form, and the airflow velocity within the building is accelerated. The orientation of the structure also determines whether the inner U space is protected from the airflow.

Fourth, L-shaped buildings create patterns of air movement similar to those of linear forms (Figure 6.11). The area of calm created by the building and the degree of potential airflow through the building are deter-

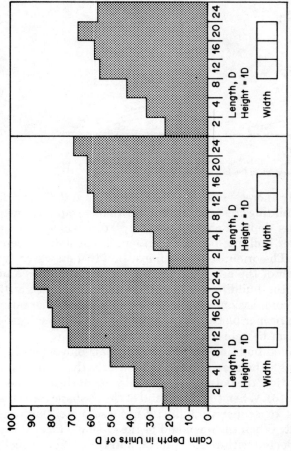

Figure 6.8 The size of the calm increases in length and depth with increase in building length. The rate and amount of increase are not as great as the increase in building length. [1]

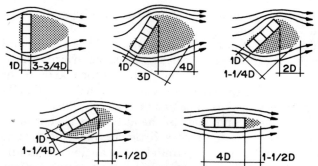

Figure 6.9 A linear building form creates varying sizes of protected areas depending on the direction of the air movement. [1]

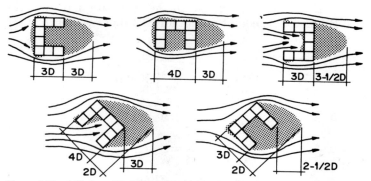

Figure 6.10 A U-shaped building form creates relatively the same size of protected area regardless of its orientation. However, the orientation does affect the potential for maximum air movement through the building and determines whether the inner U is protected. [1]

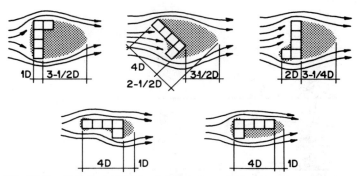

Figure 6.11 An L-shaped building form is similar to the linear form in that its calm area varies in size with the form's orientation to the airflow. [1]

mined by whether the short or long side faces the airflow. The greatest potential for optimum air movement within the structure occurs when the building is angled 45° to the oncoming airflow.

Fifth, T-shaped building forms are a combination of the square and linear forms. Although the calm area is basically the same in size regardless of building orientation (Figure 6.12), the potential for maximum or optimum air movement within the structure varies extensively. The T-shaped building form is best visualized as two intersecting linear forms. When the stem of the T is perpendicular to the airflow, it is ventilated thoroughly while the crossbar experiences little air movement. The reverse is true when the stem of the T is parallel to the airflow. Optimum air movement occurs when the form is skewed to the oncoming airflow, especially when the T shape catches the air.

Although the configuration and orientation determine the length, depth, and shape of the eddy, the height of the structure may alter the depth (Figure 6.13). Building height cannot affect either the length of the

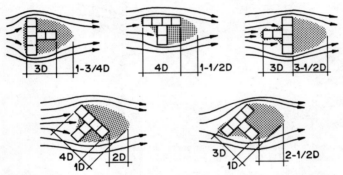

Figure 6.12   A T-shaped building form creates roughly the same size calm area regardless of its orientation. [1]

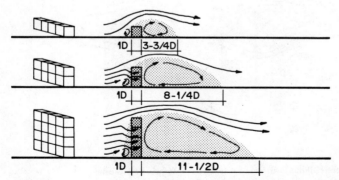

Figure 6.13   A tall structure creates a larger calm area than a comparatively smaller structure. [1]

eddy, since the length of the building is unchanged, or the shape of the calm area, because the form of the structure has not been altered. However, eddy depth may be increased as the structure is increased in height and enlarges the jump of the airflow over its top. Because of the longer jump, the pattern of the air movement is increased but not otherwise changed.

Greater building height has other side effects. The air patterns within the spaces of the building are altered, and the energy requirements of the structure are modified. As the building is increased in height, the distribution of airflow paths about it changes. More air passes around its sides, and a third of the air travels over it. That causes less upward air movement along the windward face except at the top, where an upward flow of air is always present. As a result, the top floor or the top one-third floors of a building experience an upward air movement within the structure's interior spaces (Figure 6.14).

As building height causes changes in air movement patterns, alteration of the energy demand of the structure also occur. Although a two-story building has a smaller surface exposed to the elements than a single-story building, it is not as easy to protect from those elements. As a structure is made greater in height, it is less likely to be protected from undesirable air movement, sheltered from direct solar heat gain, ventilated at the proper height within interior spaces, and benefited by precise air movement control. In addition, the stack effect in multiple-story buildings makes controlling warm and cool air at all levels more difficult. That is true in both naturally ventilated and air-conditioned structures. Consequently, increasing building height should be avoided in the desire to achieve comfortable living spaces and optimum air movement within those spaces.

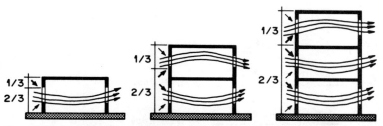

Figure 6.14 Pressure differentials on the windward side of a building are altered as the structure is increased in height; consequently, the pattern of air movement within the building is modified.

Roof forms not only affect exterior airflow patterns but also influence air movement within the structure. The slope of a roof, or the lack of one, determines the volume, velocity, and pattern of the airflow over the building and through its interior. As the roof slope is increased, the volume of air flowing over the top of the building becomes greater (Figures 6.15, 6.16 and 6.17); consequently, less air movement is available for use within the structure. The size of the eddy, with regard to roof forms, appears to be a product of roof slope rather than an expression of potential air movement through the structure.

The jump, or the area of calm, is caused by the increased height of the building as well as the slope. A shed roof can illustrate the point. When the low edge of the shed roof is on the windward side of the building and the roof slope is increased, little effect on the eddy occurs (Figure 6.18).

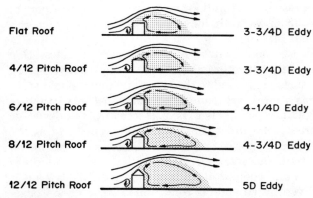

Figure 6.15 As the roof slope is increased, the area of calm gains size in both height and depth, especially the latter. [1]

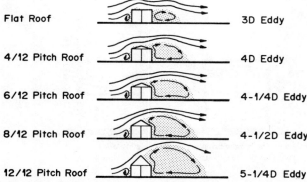

Figure 6.16 The eddy is increased equally in height and depth as the slope of the roof is made steeper. [1]

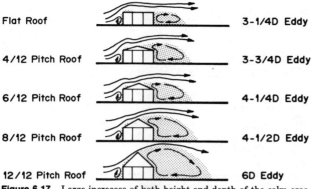

Flat Roof		3-1/4D Eddy
4/12 Pitch Roof		3-3/4D Eddy
6/12 Pitch Roof		4-1/4D Eddy
8/12 Pitch Roof		4-1/2D Eddy
12/12 Pitch Roof		6D Eddy

Figure 6.17 Large increases of both height and depth of the calm area occur as the slope of the roof is made steeper. [1]

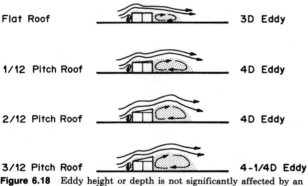

Flat Roof		3D Eddy
1/12 Pitch Roof		4D Eddy
2/12 Pitch Roof		4D Eddy
3/12 Pitch Roof		4-1/4D Eddy

Figure 6.18 Eddy height or depth is not significantly affected by an increase of roof slope. [1]

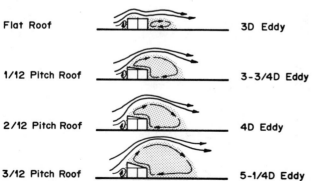

Flat Roof		3D Eddy
1/12 Pitch Roof		3-3/4D Eddy
2/12 Pitch Roof		4D Eddy
3/12 Pitch Roof		5-1/4D Eddy

Figure 6.19 The increase of roof slope increases both the height anddepth of the calm area. [1]

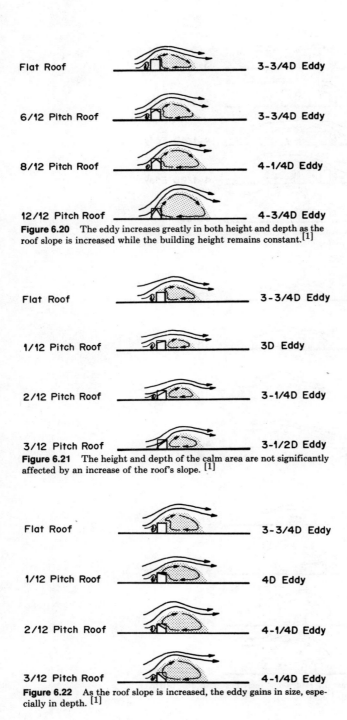

Flat Roof   3-3/4D Eddy

6/12 Pitch Roof   3-3/4D Eddy

8/12 Pitch Roof   4-1/4D Eddy

12/12 Pitch Roof   4-3/4D Eddy

Figure 6.20 The eddy increases greatly in both height and depth as the roof slope is increased while the building height remains constant.[1]

Flat Roof   3-3/4D Eddy

1/12 Pitch Roof   3D Eddy

2/12 Pitch Roof   3-1/4D Eddy

3/12 Pitch Roof   3-1/2D Eddy

Figure 6.21 The height and depth of the calm area are not significantly affected by an increase of the roof's slope. [1]

Flat Roof   3-3/4D Eddy

1/12 Pitch Roof   4D Eddy

2/12 Pitch Roof   4-1/4D Eddy

3/12 Pitch Roof   4-1/4D Eddy

Figure 6.22 As the roof slope is increased, the eddy gains in size, especially in depth. [1]

On the other hand, when the low edge of the shed roof is on the leeward side of the building and as the roof slope is increased, the calm area is enlarged greatly in both height and depth (Figure 6.19). The change of eddy size appears to be more of a product of building height and a direct obstruction of airflow rather than roof slope alone. The best way to determine whether that is true is to compare structures with different degrees of roof slopes while keeping building height constant.

As the slope of a gable roof form is increased in pitch, the calm area constantly increases as a direct product of the slope as it deflects more and more of the airflow (Figure 6.20). A shed roof form with the low edge of the windward side of the building has little effect on eddy size regardless of roof slope (Figure 6.21). However, when the low edge of a shed roof is on the leeward side of the building, the eddy will gain some depth, but the height remains essentially the same (Figure 6.22). Although both building height and roof slope affect the airflow, the height of the building appears to be the major factor determining eddy size.

Calm areas located in the immediate vicinity of the building form are not usually the product of a single element; instead, they are the result of the interaction of several elements. The slope of the roof form seems to determine the height of the calm area and, indirectly, influence the depth of the eddy. In addition, the air movement within the structure is determined by the size of the positive pressure eddy on the windward side of the building and the size of the negative pressure eddy on the leeward side.

The windward calm area determines the quantity of air which will enter the building, and the leeward eddy zone determines the velocity of the airflow through the structure. As the windward eddy increases in size, the volume of air introduced into the structure becomes greater. As the leeward eddy enlarges, the speed of the air movement increases, since negative pressure on the leeward side of the building is greater. The roof form and slope, along with building configuration and orientation, extensively influence air movement quantities, velocities, and patterns in an intertwined relationship.

The impact of roof forms does not stop at the edge of the building's wall; it extends on into overhangs. Such structural projections are useful for their ability to deflect more air into the interior spaces of a building. Although the leeward eddies of structures are not significantly altered by them, overhangs do increase the quantity and velocity of air movement through the building when the projection is on the windward side of the structure. This phenomenon is a result of the extension's ability to capture the movement of air and direct it into the building.

Overhangs on a flat roof building do not allow the upper airflow to travel over the top of the structure (Figure 6.23). When the projection is on the leeward side of the building, only a slight variation of the eddy occurs as the eddy gains a minute extension of depth (Figure 6.24). The volume and speed of the airflow within the exterior spaces of the structure are not

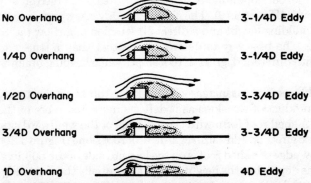

Figure 6.23 The overhang is increased in depth, the windward eddy gains size steadily and the leeward eddy increases slightly. [1]

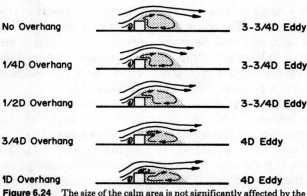

Figure 6.24 The size of the calm area is not significantly affected by the extension of the projection. [1]

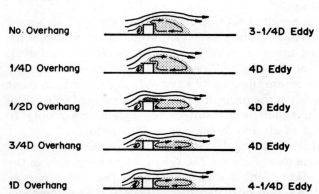

Figure 6.25 The windward eddy enlarges constantly and the leeward eddy remains constant in size as both overhangs are increased in depth. [1]

significantly increased. Even when overhangs are extended on both the windward and leeward sides of the building, the results are relatively the same as when only the windward projection is present (Figure 6.25).

A gable roof with extensions on the windward and leeward sides of the structure creates no appreciable changes in the leeward eddy (Figure 6.26).

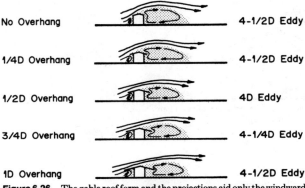

No Overhang	4-1/2D Eddy
1/4D Overhang	4-1/2D Eddy
1/2D Overhang	4D Eddy
3/4D Overhang	4-1/4D Eddy
1D Overhang	4-1/2D Eddy

Figure 6.26 The gable roof form and the projections aid only the windward calm area to gain in size while the leeward calm area remains unchanged. [1]

It has the identical effect on air movement as a windward projection on a flat roof building. The shed roof form involves a different phenomenon in comparison with other roof forms and their overhangs (Figure 6.27). As the projection is made deeper on the high edge, it also increases in height as it is extended outward and upward. When the overhang is on the high edge and windward side of the building, both windward and leeward eddies gain in volume (Figure 6.28). Consequently, an increased quantity and accelerated velocity of air movement may occur within the structure as the projection captures the oncoming airflow. Overhangs accentuate the building's ability to guide, obstruct, and divert air movement within the immediate environment of the structure.

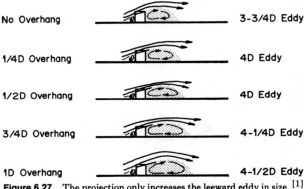

No Overhang	3-3/4D Eddy
1/4D Overhang	4D Eddy
1/2D Overhang	4D Eddy
3/4D Overhang	4-1/4D Eddy
1D Overhang	4-1/2D Eddy

Figure 6.27 The projection only increases the leeward eddy in size. [1]

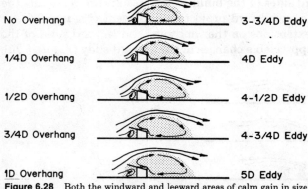

No Overhang		3-3/4D Eddy
1/4D Overhang		4D Eddy
1/2D Overhang		4-1/2D Eddy
3/4D Overhang		4-3/4D Eddy
1D Overhang		5D Eddy

Figure 6.28 Both the windward and leeward areas of calm gain in size as the overhang is extended further in depth. [1]

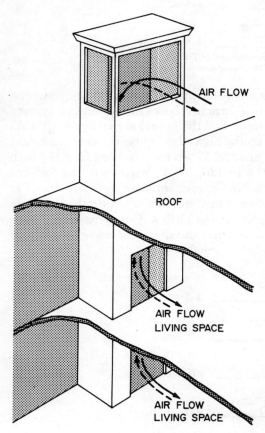

AIR FLOW

ROOF

AIR FLOW
LIVING SPACE

AIR FLOW
LIVING SPACE

Figure 6.29 The wind tower operates by either or both pressure differentials and buoyancy. *(Adapted from Ref. 2.)*

In addition to roof forms and overhangs, other architectural elements have been used to control air movement; wind towers, breezeways, and courtyards may or may not enhance the potential of airflow within structures. Wind towers capture the airflow and guide it down the tower and into interior spaces below. The forces of pressure differentials of the moving air drive the air down the tower shaft (Figures 6.29 and 6.30).

When no air movement is present, the tower will establish air movement within its shaft by altering the temperature of the air. Consequently, the air density will change in and around the tower. "The difference in density creates a draft, pulling air either up or down through the tower."[2] Openings in the living spaces allow air to enter or exit those spaces and thereby create air movement in the structure.

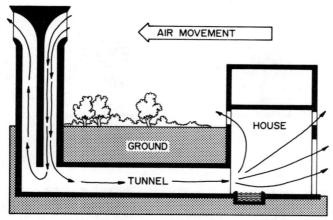

Figure 6.30 Air movement forces air down the wind tower on the windward side and up on the leeward side. Buoyancy may cause either a rise or fall of air within the tower depending on whether the inside air or outside air is cooler. *(Adapted from Ref. 2)*

The architectural designer must also be aware of two possible factors which may hinder the proper functioning of the wind tower system. First, the tower's design should respond to at least the major flow of air direction by readily catching the air movements which benefit the building's interior spaces. Ideally, the tower should be omnidirectional (Figure 6.31). Second, the wind tower system must create a balance between the forces of pressure and buoyancy. Density differential is a weak method of moving air, and even the slightest breeze will disrupt the effectiveness of buoyancy as an air movement force. Consequently, the wind tower should be flexible in order to satisfy the requirements of the building's occupants.

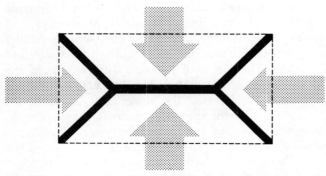

Figure 6.31 The design of the wind tower's openings should capture air movement in all directions, but especially in the predominant directions.

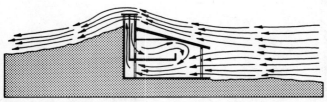

Figure 6.32 Earth-sheltered houses may utilize wind towers to aid in increased air movement within the structure.

It provides a great opportunity for thorough air movement within underground residences (Figure 6.32).

Breezeways catch, funnel, and accelerate the flow of air movement through them. However, if the flow of air enters a breezeway in a direction opposite to the expected one, the air mass may be diffused and impeded, which may or may not be a desirable effect (Figure 6.33).

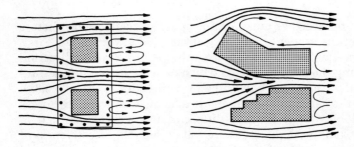

Figure 6.33 Either inlets and outlets of breezeways should be of equal size or the outlets should be larger than the inlets in order to achieve optimum air movement.

Courtyards not only control air movement precisely but create their own individual micro-climates. If properly designed, they provide a comfortable space for people. Furthermore, the surrounding interior spaces may not require mechanical cooling or heating because they receive free energy from the courtyard. The free energy may also include natural lighting. The courtyard or atrium may further function as a thermal chimney driven by the forces of buoyancy when inlet openings are supplied near the bottom of the courtyard (Figure 6.34).

The building should be constructed of thermally light materials so undesirable heat will not be stored in the courtyard walls, which will radiate back into the courtyard space and disrupt the buoyancy cycle of the

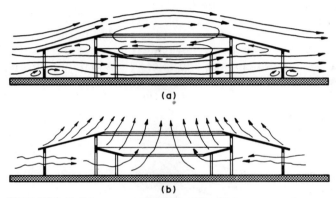

Figure 6.34 Both buoyancy and pressure differentials may cause air movement within the unique design of a courtyard. (a) Pressure drives cool air through the structure while warm air travels over the building. (b) Buoyancy creates a cycle of air movement as hot air rises from the rooftops and courtyard. Cool air moves through the interior spaces as a result of buoyancy and pressure differentials.

air movement. Trees, shrubs, and other shading devices aid in alleviating the heat load in the courtyard. On the other hand, the use of thermal mass and ventilation of cooler night air permits the storage of cold within the atrium space that will be released during the day. In short, the courtyard or atrium will "perform better as an unconditioned space if it is . . . shaded but otherwise open . . . " [3]

As buildings alter the air movement, new micro-climates are created in their surroundings. Here lies the opportunity for architectural designers to control air movement as a means of creating quality spaces rather than having the air movement determine the quality of the spaces. Along with designing the configuration, volume, size, voids and solids, height, lighting level, and other details of architectural spaces, the designer may also designate the temperature, humidity, and air movement within the spaces. These areas may be positioned within, outside, below, or above or be encompassed by the building form.

Buildings and Topography

Because buildings must stand upon the ground, the surface of the earth has an enormous impact upon them. Besides the direct effect of the topography on building configuration, the earth's landforms alter the air movement. Topography, along with temperature, humidity, sunlight, and air movement, determines the micro-climates of the world. Its contours, surface textures, and adjacency to bodies of water affect the character of the air movement.

The contours of landforms have a wide variety of influence on the movement of air. Whether at the meso-climatic or micro-climatic scale, concave and convex contours affect airflow. Contours may accelerate the airflow or slow it down, or they may sway the air movement pattern (Figure 6.35). The distinction between meso-climatic and micro-climatic landforms is that the latter may be altered by the architectural designer within certain limits. The micro-climatic contours exist within the structure of the meso-climatic landform system.

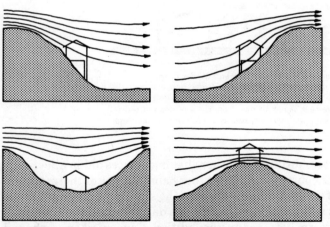

Figure 6.35 The topography may decrease or increase the velocity of the airflow as well as alter the pattern of the air movement.

The only difference between the two is scale, including the effect on air movement. Furthermore, the quality of the moving air mass is conditioned by the slopes over which the air mass travels as it either gains heat from or loses heat to the slopes. South slopes are usually warmer than flat land because they receive more direct sunlight. North slopes, on the other hand, are the coolest areas because they obtain no direct sunlight at any time of the year. East slopes are better than west slopes because they have the advantage of the morning sun without the intense heat of the afternoon sun. Therefore, it is preferable to locate residences on east or southeast slopes.

Situating the residential structure on the upper portion of a slope permits it to take advantage of prevailing air movement, which is conditioned by the quality of the slope. Such contours develop specific microclimatic environments which architectural designers may choose to utilize for their particular design needs (Figure 6.36). As nature conditions the temperature through air movement and landforms, a surplus of free energy is sitting idle—a ready resource for utilization.

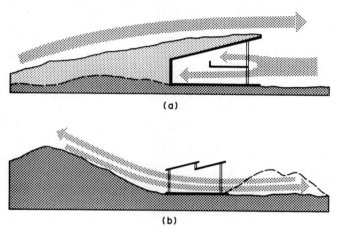

(a)

(b)

Figure 6.36 Landforms at the micro-climatic scale may be (a) added as a method of controlling desirable and undesirable air movement or (b) subtracted to gain additional air movement.

The surface textures of the ground influence air movement velocity, pattern, and quality. As the ground cover increases in coarseness, the airflow velocity decreases because of the frictional drag of the ground (Figure 6.37). For example, the speed of a given air mass will be lower as the mass travels over a sandy beach, a lawn, small plantings, a parking lot with cars, a stand of trees, and a residential area, respectively. Furthermore, "the rougher the surface of the terrain, the greater will be the vortex on the windward side of a building. This means that when the ground surface is rough instead of smooth, the air will start to flow over the structure at a greater distance away from the building." [4]

As the airflow encounters more obstructions and disturbances along the ground plane, greater variances occur within it. The quality of the air mass may vary in character over many miles, several hundred yards, or even a few feet. The degree of albedo, or reflectivity of the terrain, and conductivity determine the nature of the microclimatic environment. If the

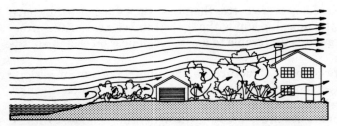

Figure 6.37 The texture of the ground surface determines the velocity and pattern of air movement. The rougher the texture the slower the velocity of airflow and the more altered the pattern of air movement.

ground's albedo is low and it conductivity is high, the microclimate is mild and stable because excess heat is collected and stored in the ground. If the ground's albedo is high and its conductivity is low, the microclimate shifts between extremes because no heat is retained in the surface.[5] As air moves over these surfaces, it loses or gains heat (Figure 6.38). To illus-

Figure 6.38 Bodies of water, grass, trees, and shrubs tend to even out the fluctuations of the microclimate, and walks, pavements, sand, and snow tend to accentuate the extreme shifts of the microclimate.

trate, temperature differences as great as 54°F have been known to occur with temperatures of 52°F on the grass and 106°F on the pavement; and even when the ambient air temperature was 77°F, the temperature of a concrete walk was 95°F and the temperature of a dark slate roof was 110°F. [5] An air mass moving over a rooftop could gain as much as 33°F. The character of air movement is the product of immediate environmental factors.

Land adjacent to bodies of water usually receives a constant motion of air. During the day, the land heats up faster than the water, and the cool air from the body of water replaces the warm air rising from the land. Consequently, a daytime movement of air comes from the body of water. Since the water retains heat longer than the land, the reverse occurs at night. Warm air rises from the body of water and is replaced by the cool air from the land. Therefore, a nighttime movement of air travels off the land and onto the body of water. A constant dynamic interaction occurs between the land and the water as these bodies shift in their relative heat content (Figure 6.39).

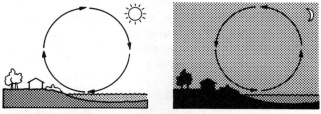

Figure 6.39 A cycle of air movement occurs between the land and a body of water. The direction of flow reverses at night.

Buildings and Vegetation

Vegetation has always played a role in the design of residences. Architectural designers are advancing beyond the inherent beauty of landscaping. They are utilizing vegetation as an energy-saving device by reducing solar radiation on buildings and controlling the conditioning air movement. Nevertheless, the function of landscaping should not override the aesthetic characteristics. The two must be compatible.

Shrubs are nature's sophisticated fences. However, unlike most fences, shrubs do not create unpredictable and violent eddies. Instead, they break the main thrust of the airflow as the air is filtered into a gentle stream (Figure 6.40). Space beneath the shrubs allows the air to circulate. The flow of air which travels over the shrubs is accelerated. Shrubs are essentially like small trees. The two have similar effects on air movement; only the scale is different.

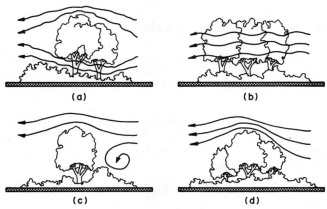

Figure 6.40 Vegetation influences the pattern of air movement through (a) guidance, (b) filtration, (c) obstruction, and (d) deflection. *(Adapted from Ref. 6)*

Trees, in regard to air movement, are large shrubs. They not only absorb sunlight and heat but also control the flow of air. The foliage of the trees filters the air movement, and the air velocity beneath the canopy is increased. The extent of airflow deflection varies with the height, width, density, arrangement, and species of vegetation. Deciduous trees direct air movement at the optimum during the summer months and to a slight degree in the winter months. Evergreens work year-round.

In addition, as trees or shrubs gain height, more air is allowed to pass under the foliage and create an increase in air movement velocity. Andre Le Notre, a famous French landscape designer of the 1600s, advised that "trees placed near the house should be kept at the same distance from the house as the height of the house." [7] Trees so spaced create excellent air movement control devices or breaks, and the problem of leaves clogging up gutters and downspouts is avoided.

Vegetation affects its immediate micro-climatic environment by blocking the sun's radiation from reaching the ground. Of the total solar radiation entering the earth's atmosphere, only an average of 5 percent passes through tree canopies (Figure 6.41). The remaining percent is utilized by the trees (42 percent) or reflected back into the atmosphere (53 percent). A single tree can absorb over a half billion Btus of solar energy a day. The net effect is lower ambient air temperatures beneath the vegetation.

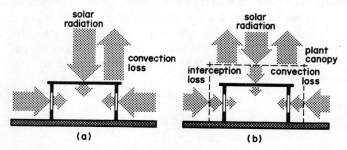

Figure 6.41   (a) An unprotected building experiences intense solar radiation and buildup of heat within the structure. (b) Plant canopies, such as those of trees, protect a building from solar radiation and dissipate absorbed heat from the structure. (*Adapted from Ref. 8*)

By controlling and conditioning air movement through the use of vegetation, architectural designers can not only create more physically comfortable and energy-efficient living spaces, but also provide greater psychological and aesthetical comfort.

Buildings and Fences

Fences are man's imitations of nature's shrubs in providing privacy, barriers, boundaries, and air movement control around buildings. They may create calm areas, establish zones of air disturbance, increase or decrease airflow, and change airflow patterns. Such effects may or may not be desirable within the perimeter of a building. Consequently, the location and design of fences need careful consideration with regard to their effect on air movement characteristics.

Michael O'Hare and Richard E. Kronauer studied and tested ten different types of fences with some quite interesting results (Figure 6.42). The test models corresponded to a 5 ft 6 in high fence with the

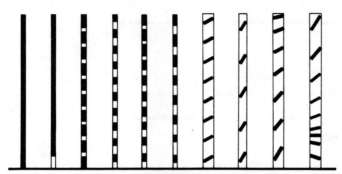

Figure 6.42   Ten different types of fences tested in a wind tunnel. Each has a different percentage of openness, slat placement, and slat angle. (*Adapted from Ref. 9*)

slats representing 2 by 10's. The airflow in the studies was perpendicular to the fences, and the following conclusions were established (Figure 6.43).

1. The solid fence created a definite negative pressure zone and reverse flow eddy immediately behind it.

2. The solid fence with a slot at the base was no significant improvement over the solid fence except for a slight reduction of the reverse flow eddy.

3. The 25 percent open horizontal-slat fence provided the lowest velocities near the fence, but its downstream performance was not as good as that of the 50 percent open horizontal-slat fence.

4. The 35 percent open horizontal-slat fence created the least reverse flow eddy, but its downstream performance was not as good as that of the 50 percent open horizontal-slat fence.

5. The 50 percent open horizontal-slat fence had excellent performance downstream while sustaining a high velocity and reverse flow eddy near it.

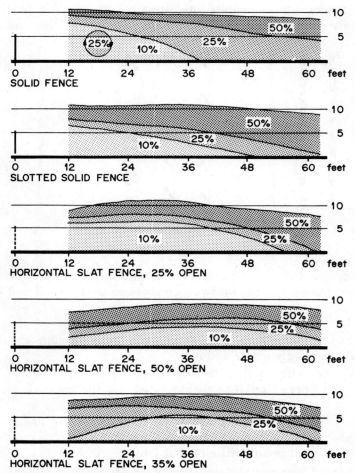

Figure 6.43 The fences of Figure 6.42 provide different airflow patterns and velocities. The velocity of the air movement ranges from 10 to 50 percent in comparison with the initial free stream velocity. *(Adapted from Ref. 9)*

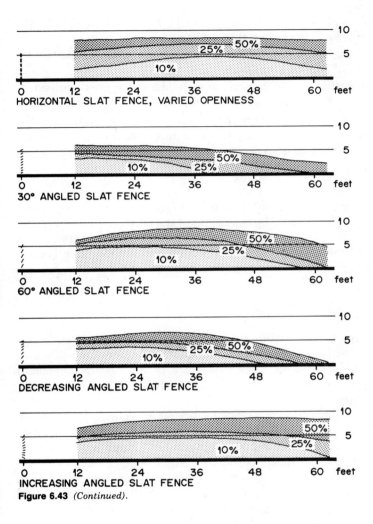

HORIZONTAL SLAT FENCE, VARIED OPENNESS

30° ANGLED SLAT FENCE

60° ANGLED SLAT FENCE

DECREASING ANGLED SLAT FENCE

INCREASING ANGLED SLAT FENCE

Figure 6.43 *(Continued).*

6. The varied open, 40 to 60 percent, horizontal-slat fence developed extremely low velocities downstream and no reverse flow eddy while maintaining a velocity near the ground of less than 20 percent of the free stream velocity.

7. The 30° angled 25 percent open slat fence produced high downstream velocities and no reverse flow eddy.

8. The 60° angled 25 percent open slat fence created results similar to those of the 30° angled slat fence.

9. The decreasing angled 25 percent open slat fence provided results resembling those of the 30° angled slat fence.

10. The increasing angled 25 percent open slat fence established the lowest velocity downstream of any of the fences tested as well as a low velocity near it and created no reverse flow eddy. In short, this fence was the best windbreak and was clearly superior. [9]

Fences, as revealed by this study, have two major characteristics with respect to air movement control. The pattern and velocity of the airflow near the fence are determined by the height of the fence, and the air movement downstream depends on the overall porosity. The shape or the configuration of the fence is immaterial with respect to the effectiveness of the fence as an air movement control device.

Buildings and Adjacent Buildings

A structure often stands within a network of other buildings, such as a downtown district or a residential neighborhood. Consequently, one building affects another in numerous ways including air movement through and about them. A building form influences its immediate environment, and vice versa.

By its position, a building induces a particular air velocity and pattern as the air deflects around it. Buildings angled to the airflow may decrease the velocity of the air as much as 50 to 60 percent. [10] Buildings positioned in a row may develop pockets of turbulence which contain little air movement that creates an unusual leapfrog airflow pattern (Figure 6.44). In addition, the spaces between buildings cause the airflow along the ground to channel into narrow streams. Consequently, succeeding buildings receive no significant air movement (Figure 6.45).

When the buildings are positioned in an alternating pattern, the flow of the air is deflected off each succeeding building as the air travels downstream (Figure 6.46). In this situation, each building receives air movement. The position of individual buildings in a group may be determined

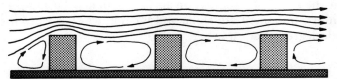

Figure 6.44 Buildings in a row create an area of calm downstream which affects subsequent buildings. *(Adapted from Ref. 10)*

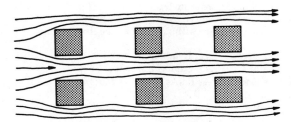

Figure 6.45 Linearly arranged buildings protect or block subsequent buildings from potential airflow. *(Adapted from Ref. 10)*

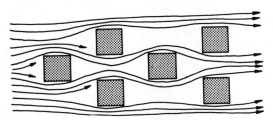

Figure 6.46 Alternating buildings linearly actually enhances the potential of airflow around each building. *(Adapted from Ref. 10)*

by whether air movement is desired in a given direction or directions. The airflow velocity and pattern are affected by the positions of various buildings. The dynamic flow of air movement becomes more complex as the relation of buildings to one another are varied.

As buildings alter the pattern and velocity of the airflow, the immediate environment and surrounding buildings are affected. One example is the John Hancock Building in Boston, Massachusetts. Unusual air movement patterns and velocities created by both the existing adjacent structures and the new structure contributed to the popping out of windows.

The control of air movement with regard to changing patterns and varying velocities is extremely intricate. Engineers have established techniques of determining its effects on structures to a mathematical and scientific level, but the effect of buildings on air is still an art. Air movement control is an art, but there are skills and techniques that can be applied with a high degree of accuracy.

REFERENCES

1. Benjamin H. Evans, *Natural Air Flow Around Buildings*, Research Report No. 59, Texas Engineering Experiment Station, College Station, Tex., 1957.
2. Mehdi N. Bahadori, "Passive Cooling Systems in Iranian Architecture," *Scientific American*, February 1978.
3. Donald Watson, "The Energy within the Space within," *Progressive Architecture*, vol. 63, July 1982.
4. Jeffrey Ellis Aronin, *Climate and Architecture*, Reinhold, New York, 1953.
5. Kevin Lynch, *Site Planning*, The M.I.T. Press, Cambridge, Mass., 1971.
6. Carol A. Smyser and the editors of Rodale Press Books, *Nature's Design*, Rodale Press, Emmaus, Pa., 1982.
7. James D. Higson, *Building and Remodeling for Energy Savings,*, Craftsman Book, Solana Beach, Calif., 1977.
8. Stephen Sesiuk, "Architectural and Environmental Horticulture: An Investigation into the Use of Vegetation for Energy Conservation," *Environmental Design: Research, Theory, and Application,* Vol. 10, Environmental Design Research Association, Washington, D.C., 1979.
9. Michael O'Hare and Richard E. Kronauer, "Fence Designs To Keep Wind From Being a Nuisance," *Architectural Record,* Vol. 146, July 1969.
10. Victor Olgyay, *Design with Climate,* Princeton University Press, Princeton, New Jersey, 1963.

Chapter

7

Building Openings

Of necessity, buildings have openings to provide for access and also for natural lighting, ventilation, and visual release. A building with well-located openings would be more effective in conserving energy than one with almost no openings at all. That is especially true when air movement is used to create human comfort within.

Air movement within a building is affected by the orientation, size, placement, ratio, and types of openings, which alter the inertia, pressure differentials, and buoyancy characteristics of airflow. The openings may guide the air into specific patterns and regulate its velocities. Several techniques of air movement control are disclosed by examining the interaction of building openings and air movement.

Placement and Orientation

Air is deflected over and around building forms and thereby creates the zones of pressure differentials known as calm areas or eddies. They are positive and negative pressure zones (Figure 7.1). Positive pressure is exerted on the windward side of a building as the air piles up. This zone causes the airflow to slow down until a new path is found around the obstacle. Once a new release is located, the air speeds around the obstacle at a velocity greater than that of the initial airflow. Likewise, negative pressure is formed on the leeward face and sides of the building because of lower air density. These zones are potentially optimum locations for openings in the building's skin to encourage air movement through the structure.

Inlet openings should be located in positive pressure zones and outlet openings in negative pressure zones. That provides the best conditions for maximum air movement through the building (Figure 7.2). However, the essential function of air movement within the building is to provide comfort for the occupants, and it is essential that the occupants benefit directly.

Section   **Plan**

Figure 7.1   The movement of air over a structure creates positive and negative pressure zones.

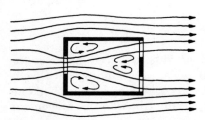

Figure 7.2   Air flows from positive to negative pressure zones through a building. The inlet opening is located in the positive zone and the outlet openings are located in the negative zone in order to maximize air movement within the building.

To illustrate, air movement should occur at the standing and sitting levels in a living room and at the sleeping level in a bedroom. In the dining room, the flow should occur around the sitting level for personal cooling, but it should not cool the food on the table. Therefore, the air needs to travel at the level of a sitting person's head and shoulders but not at the table's height. The flow may also occur at the foot and knee level.

A balance between maximum air movement, human comfort, and function of the spaces is required. The equilibrium between those factors is ever-changing. Consequently, flexible control of air movement may be desirable within cost limitations.

The placement of openings in relation to one another and the orientation of the openings to the direction of airflow may create positive and/or negative results. Contrary to popular belief, optimum air movement through a space does not always occur when the inlet openings directly face the source of the airflow. "In some cases, better conditions can be achieved when the wind is oblique to the inlet windows." [1]

When the single openings are on opposite sides of an interior space and the air movement is perpendicular to the inlet opening, the main airflow travels from inlet opening to outlet opening (Figure 7.3). The remainder

Figure 7.3 (*a*) When air movement is perpendicular to the inlet opening and is aligned with both openings, the flow of air will pass through the interior space in a narrow stream. (*b*) When air movement is oblique to aligned inlet and outlet openings, the flow of air will circulate throughout the entire interior space.

of the interior space receives no significant air movement. When the air movement is skewed to the plane of the inlet opening, most of the airflow takes up "a turbulent, circling motion around the room, increasing the airflow along the side walls and in the corners." [1] On the other hand, when the openings are on adjacent sides of an interior space and the air movement is skewed to the inlet opening but aligned with both openings, the main airflow traverses directly from the inlet opening to the outlet opening (Figure 7.4). When the air movement is perpendicular to the inlet opening, turbulence occurs within the interior space and creates circular currents encompassing the entire space. Unless the occupants are sitting in the narrow stream of air movement, the interior space with the turbulence provides better overall air movement or ventilation.

 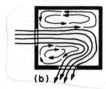

Figure 7.4 (*a*) When air movement is skewed to the inlet opening but the inlet and outlet openings are in alignment with the exterior direction of the air movement, the flow of air will pass through the interior space in a narrow stream. (*b*) When the air movement is perpendicular to the inlet opening and the outlet opening is in an adjacent wall, the flow of air will circulate throughout the entire interior space.

Air movement continues in its initial direction until it encounters an obstruction, loses momentum because of friction, engages with buoyancy forces, or impacts pressure differentials. In other words, air movement has inertia. Because of those dynamic forces, the flow of air does not always take the shortest path between two points, such as an inlet opening to an outlet opening. Pressure differentials near the inlet may alter the pattern and velocity of the airflow. When the pressures are symmetrical around the inlet opening, the flow of air is perpendicular to the plane of the opening; when they are asymmetrical, the airflow may be at any angle other than perpendicular (Figure 7.5). The change in the flow of the air may occur in either the horizontal or vertical plane.

(a)   (b)

Figure 7.5 Varying pressures around an inlet opening create different airflow patterns. The flow is straight when the pressures are symmetrical (a) and skewed when they are asymmetrical (b).

Buoyancy, like pressure differentials, may vary the airflow but only in minor degrees, since buoyancy is weaker than pressure differentials . A cooler interior space may cause the airflow to dip downward, whereas a warmer interior space may cause the airflow to bend upward as the flow of air respectively decreases or increases in heat content (Figure 7.6). However, if the airflow remains relatively constant in temperature, the reverse is likely to occur. The important issue to remember is that any change in the airflow pattern extracts energy from and reduces the velocity of the airflow.

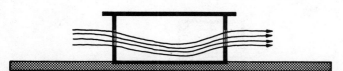

Figure 7.6 Buoyancy may cause a dip in the airflow traveling through an interior space because of an interior-to-exterior thermal differential.

Size and Ratio

Openings that allow air to enter the space (inlets) and openings that permit air to exit the same space (outlets) are the ventilating devices of the building. An inlet may become an outlet, and vice versa, depending on the direction of the airflow. The openings may be windows, doors, vents or ventilators, and/or specially designed openings. Inlet openings, which are usually located in positive pressure zones, play a dominant role in determining the pattern of the air movement. Outlet openings, which are most often found in negative pressure areas, regulate the airflow velocity. The key elements are the size of the openings and their relations to each other.

The size of the openings has a substantial effect on interior airflow velocities if both an entry and exit are provided. If there is only one opening, its size has little effect on airflow velocity (Figure 7.7). However, when the airflow is oblique to the inlet opening, greater variations in the air pressure along the sides of the interior space occur that allow air to enter one part of the opening and exit through another. [1] The activity of air movement within the space accelerates as the opening size is increased. On the other hand, when the airflow is perpendicular to the inlet opening, pressure differences along the sides of the interior space are too small for the increase in opening size to create significant increased air movement. [1]

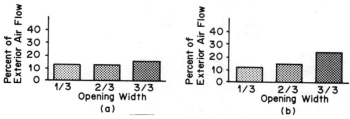

Figure 7.7 The percentage of air movement within an interior space compared to the exterior airflow appears to be relatively unaffected, especially when the flow of air is perpendicular to the inlet opening (a). When the airflow is skewed to the inlet opening and the opening size is large, the interior air movement is significantly increased (b). *(Adapted from Ref. 1)*

Interior airflow velocities are greatly increased by the mere presence of both an inlet and an outlet opening (Figure 7.8). The interior movement of air increases an average of 32 to 65 percent and to a maximum of 152 percent relative to the exterior airflow. A single opening in a space, unlike two openings, usually permits an average of 12 to 23 percent. As the inlet and outlet openings are increased in size, the velocity of the airflow does

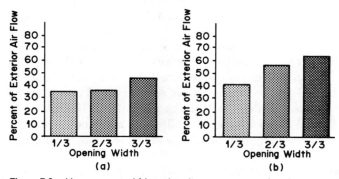

Figure 7.8 Air movement within an interior space, compared to the exterior airflow, does not increase in velocity proportionately to an increase of the opening's size. The increased velocity of the airflow perpendicular to the inlet opening (a) is less than the increased airflow skewed to the inlet opening (b). *(Adapted from Ref. 1)*

not increase proportionately. The inlet and outlet openings do not function as a consistent team in obtaining maximum airflow through an interior space, but their effects result in a compromise of forces in achieving optimum air movement.

The relationship between an inlet and an outlet opening needs to be compatible if the two are to function together. Depending on the results desired, the importance of the inlet-to-outlet ratio varies (Figure 7.9). The

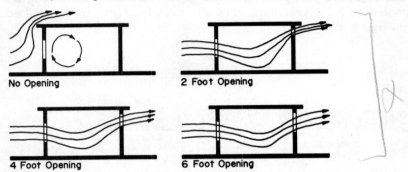

No Opening   2 Foot Opening

4 Foot Opening   6 Foot Opening

Figure 7.9   Before air movement in an interior space can be significantly realized, there must be more than one opening. As the outlet opening is increased in size, the interior airflow gains in velocity. The figures indicate the relative airspeeds in terms of percentages of the exterior air velocity. *(Adapted from Ref. 2)*

average indoor air movement velocity is determined by the size of the smaller opening, whether inlet or outlet. The maximum volume of air which can flow through a space is governed by the square footage of the smaller opening and the initial exterior airflow (Figure 7.10). The principle is similar to that of air-conditioning ducts and registers.

The ratio of the inlet opening to the outlet opening regulates the main airstream velocities (Figure 7.11). The speed of the airflow becomes greater as the ratio increases. The stream of the air movement is concentrated to a small area of the interior space rather than distributed throughout the space. The direction or velocity of the airflow through an interior space is not solely controlled by the inlet or outlet openings. Instead, it is controlled by the interaction of several factors such as the initial path of the exterior airflow and the size and ratio of the openings.

The openings in building surfaces that allow air movement to occur within the building vary widely in type, size, and shape. Also, the method of operation of each type influences the airflow pattern and velocity. The art of air movement control becomes more complex as the micro-climatic region of building openings is examined.

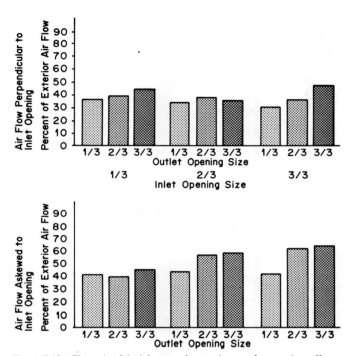

Figure 7.10 The ratio of the inlet to outlet openings produces various effects on the airflow velocities within interior spaces. As the size of the outlet increases, the airflow velocity in the interior space becomes greater. Whether the inlet and outlet openings are the same size or the inlet opening is larger than the outlet opening, the resultant interior airflow is relatively equal in both situations. For example, the 3/3 inlet with an airflow perpendicular to the inlet opening and a 2/3 outlet has a smaller percentage, 36 percent, than the 2/3 inlet and 2/3 outlet, 37 percent. The figures 1/3, 2/3, and 3/3 represent the width of the opening in proportion to the wall's width. The openings are all the same height, floor to ceiling. *(Adapted from Ref. 1)*

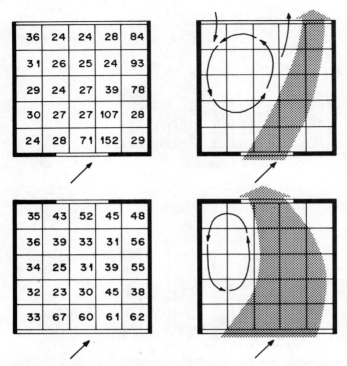

Figure 7.11 The ratio of inlet to outlet openings creates some interesting results with respect to air movement. The average indoor air movement for both types of spaces is relatively equal, 44 percent and 42 percent (*a*) and (*c*), since the smaller opening is the same for each situation. The velocity of the main airflow is different for each space as determined by the inlet-to-outlet ratio, 1:3 and 3:1, (*b*) and (*d*). (*Adapted from Ref. 1*)

Types

For summer ventilation, openings should permit maximum airflow into interior spaces for cooling. The airflow should be directed downward into the living zones. In the winter, the airflow should be guided upward to provide a cooling sensation without a draft. These are the basic ventilation functions of properly designed openings such as windows, doors, skylights, dormers, clerestories, vents, ventilators, and others of special design. This section will illustrate that some opening types are limited in their ability to facilitate optimum air movement conditions while others actually encourage and enhance the air movement.

Diagrams are utilized to illustrate air movement through openings. The eddies, patterns, directions, and velocities of the airflow are explored within each diagram.

Relatively straight arrows (→) represent the main flow of the airstream, and small wavy arrows (⤳) indicate slow air movement which is inadequate in providing human comfort. The relationships of the lines show the variations in air movement patterns, velocities, and pressure differentials. When the lines are close together, the velocity of the airflow is high and the air pressure is low or negative pressure. When the lines are spread apart, the air velocity is low and the air pressure is high or positive pressure.

Windows

The important aspects of window designs include proper daylighting, good weatherproofing, structural strength, suitable solar heat rejection or absorption, easy operation, durable construction and parts, excellent airtightness, good insulation, troublefree cleanability and serviceability, and lastly, valuable air movement control. However, the advertising literature of window manufacturers indicates a tremendous concern for all the above issues except air movement control. The fact that most window styles do not meet the requirement of air movement control further indicates lack of attention. The purpose of the examination of window styles is not to discredit any particular style, but to compare the styles in regard to air movement control. "Well-designed windows are a net gain for energy conservation; poorly designed ones are an energy burden." [3]

Virtually every one of the wide variety of window styles and sizes can be classified as one or a combination of three primary window types: simple opening, vertical-vane opening, and horizontal-vane opening. The *simple opening* includes the single-hung, double-hung, and horizontal-sliding windows that are defined as any window which opens by sliding in a single plane (Figure 7.12). The *vertical-vane opening* includes such styles as the side-hinged casement, folding casement, and vertical pivot,

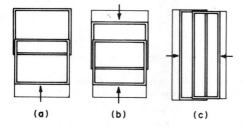

(a)   (b)   (c)

Figure 7.12 The simple opening is one type of window. Styles within this type are (a) single hung, (b) double hung, and (c) horizontal sliding.

which opens by pivoting on a vertical axis (Figure 7.13). The *horizontal-vane opening* is comprised of the projected sash, awning, basement, hopper, horizontal-pivot, and jalousie windows, which open by pivoting on a horizontal axis (Figure 7.14). However, a particular window style may fit into two or three classifications; e.g., Andersen makes a double-hung window which operates in the typical simple opening mode and switches to a horizontal-vane opening for easy cleaning and servicing.

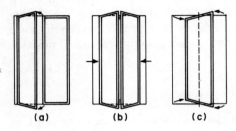

(a)   (b)   (c)

Figure 7.13 The side-hinged case-ment (a), folding casement (b), and vertical pivot (c) are styles within the vertical-vane opening type of window.

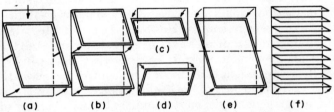

(a)   (b)   (d)   (e)   (f)

Figure 7.14 The projected sash (a), awning (b), basement (c), hopper (d), horizontal pivot (e), and jalousie (f) are styles within the horizontal-vane opening type of window.

Simple openings do not generally affect the pattern or velocity of airflow (Table 7.1). Since the windows do not project into the interior or exterior space, they do not influence the path of the incoming airstream, which continues in its initial direction. The airspeed is altered only near the window as the airstream squeezes through the opening. A double-hung window allows selection of the height of the airflow, and a horizontal-sliding window designates the placement of the airstream within the interior space. (See Figures 7.15 to 7.28.)

TABLE 7.1 Simple Openings

Single-Hung Window	
	The window may open to a maximum of 50 percent by operation of the lower sash only.
	The airstream enters the opening and continues in the same direction as the exterior airflow.
	The airstream remains at the same horizontal height as the opening.
	Adjustments of the lower sash determine the volume of air which enters the opening.

Double-Hung Window

The window has a maximum opening of 50 percent, which may be achieved with one sash open completely or two sashes open halfway.

The airstream continues in the same path as the initial exterior airflow after passing through the opening.

Through adjustments of the sashes, the airstream may enter the interior space at various horizontal levels from the top to the bottom of the window.

The airflow stays at the same horizontal height as the opening or openings.

When both sashes are adjusted to create openings at the top and bottom of the window, the airstreams enter both openings and travel parallel to each other.

When both sashes are adjusted to create equal and maximum openings at the top and bottom of the window, the airflow enters both openings simultaneously.

When both sashes are adjusted to create equal and minimum openings at the top and bottom of the window, the airflow enters both openings simultaneously.

Horizontal-Sliding Window

The window may open to a maximum of 50 percent by either one sash completely open or two sashes halfway open.

The direction of the exterior airflow remains constant as the airstream passes through the opening.

The airstream stays at the same horizontal height as the opening.

Adjustments of the sashes determine the placement of the airflow from one side of the window to the other side.

When both sashes are adjusted to create openings at both sides, the airstreams enter both openings and travel parallel to each other.

When both sashes are adjusted to create equal and maximum openings at both sides, the airflow enters both openings simultaneously.

When both sashes are adjusted to create equal and minimum openings at both sides, the airflow enters both openings simultaneously.

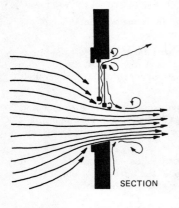

SECTION

Figure 7.15 When the single-hung window has its sash fully open, the flow of air through the opening is horizontal because of the window's central location in the wall. The slight airflow upward and into the interior space offers little value with respect to air movement within the space and human comfort. *(Adapted from Ref. 4)*

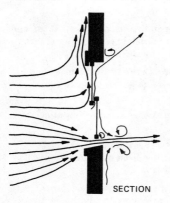

SECTION

Figure 7.16 The sash of the single-hung window is open only slightly. The airflow traveling through the opening is approximately horizontal. The minute airflow between the sashes and upward has little cooling value. *(Adapted from Ref. 4)*

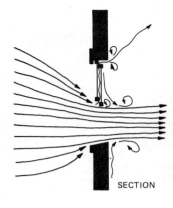

SECTION

Figure 7.17 The lower sash of the double-hung window is open completely. The airflow through the opening is horizontal because of the central location of the window in the wall. The minute upward flow of air provides little value as far as air movement within the interior space is concerned. *(Adapted from Ref. 4)*

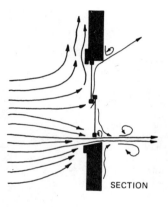

SECTION

Figure 7.18 The lower sash of the double-hung window is open only slightly. The airflow through the opening is relatively horizontal. The small airflow between the sashes and upward into the interior space offers little cooling value. *(Adapted from Ref. 4)*

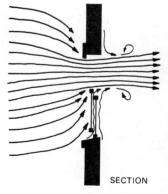

SECTION

Figure 7.19 The upper sash of the double-hung window is fully open. The air pattern is similar to that when the lower sash is open, except that the horizontal airflow is higher. The air movement between the sashes is brought into the main airstream. This opening is ideal for mixing cold fresh air with hot stale air on cold windy days. Also, on windy days, the strong airflow may enter the interior space above the occupants' heads and provide comfortable air movement. *(Adapted from Ref. 4)*

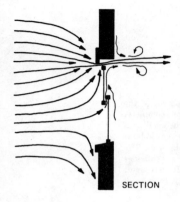

Figure 7.20 The upper sash of the double-hung window is open only slightly. The airflow through the opening is approximately horizontal. The airflow between the sashes is brought into the major airflow. This positioning of the sash may serve to reduce moisture condensation in the interior space. *(Adapted from Ref. 4)*

SECTION

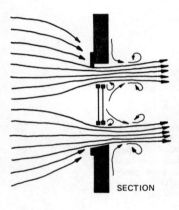

Figure 7.21 Both sashes of the double-hung window are halfway open. The exterior airflow enters both openings relatively equally, and the two airstreams are approximately parallel. The lower airflow may directly benefit the occupants; the upper stream prevents the higher air in the interior space from becoming stale. Air may enter one opening and exit from the other when cross-ventilation is not present. *(Adapted from Ref. 4)*

SECTION

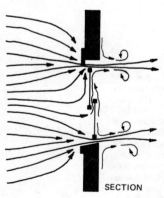

Figure 7.22 Both sashes of the double-hung window are open only slightly. The exterior airflow enters the two openings relatively equally, and the two airstreams are approximately parallel. The lower airflow may directly benefit the occupants, while the upper stream prevents the higher air in the interior space from becoming stale. Air may enter one opening and exit from the other when cross-ventilation is not present. *(Adapted from Ref. 4)*

SECTION

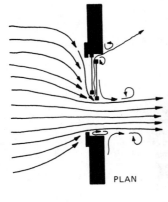

PLAN

Figure 7.23 The right-hand sash of the horizontal-sliding window is fully open. The interior airstream continues in the same path as the exterior airflow. The air movement between the sashes provides little cooling value. *(Adapted from Ref. 4)*

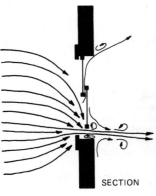

SECTION

Figure 7.24 The right-hand sash of the horizontal-sliding window is only open slightly. The interior airflow continues in the same direction as the exterior airflow. The airstream between the sashes offers little cooling value. *(Adapted from Ref. 4)*

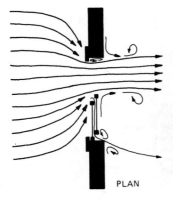

PLAN

Figure 7.25 The left-hand sash of the horizontal-sliding window is completely open. The interior airstream follows the same direction as the exterior airflow. The air movement between the sashes is brought into the main airstream. *(Adapted from Ref. 4)*

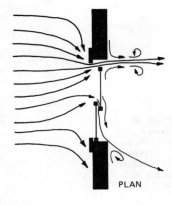

Figure 7.26 The left-hand sash of the horizontal-sliding window is open only slightly. The interior airflow continues in the same path as the exterior airflow. The movement of air between the sashes is brought into the main airstream. *(adapted from Ref. 4)*

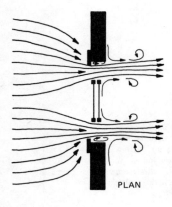

Figure 7.27 Both sashes of the horizontal-sliding window are open halfway. The exterior airflow enters both openings relatively equally, and the two airstreams are approximately parallel. Air may enter one opening and exit the other when cross-ventilation is not present. *(Adapted from Ref. 4)*

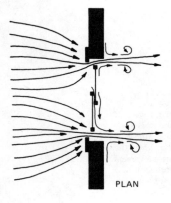

Figure 7.28 Both sashes of the horizontal-sliding window are open only slightly. The exterior airflow enters the two openings relatively equally, and the two airstreams are approximately parallel. Air may enter one opening and exit the other when cross-ventilation is not present. *(Adapted from Ref. 4)*

Vertical-vane openings exert a wide variety of influences on both the pattern and velocity of the airflow (Table 7.2). This type of opening has a design capability to control the horizontal airflow pattern. Both styles of casement windows form "a vertical 'triangular duct' shape" when the window sashes are opened to 30° angles creating three airstreams: one travels into the interior space over the window's sashes; one moves under the sashes; and the other flows through the 'triangular duct.'[4] A slight movement of window location in the wall, even a few inches, varies the pattern of the airflow. "The direction of flow is downward when the window is placed below a central position on the wall and upward when the window is placed above the central position." [4] The folding casement

TABLE 7.2   Vertical-Vane Openings

Side-hinged Casement Window

The window may open to a maximum of 100 percent.

When one sash or both sashes are open to 90°, the airflow enters the interior space at a horizontal level.

When both sashes are open only slightly, the main airstreams flow through the openings at each side.

When the exterior airflow is 90° to the window plane and either one of the sashes is open to 90°, the airflow is directed into the interior space horizontally and straight.

When the exterior airflow is 90° to the window plane and either one of the sashes is open to any angle less than 90°, the airflow is directed into the interior space at the same angle as the angled sash.

When the exterior airflow is 45° to the window plane and either one of the sashes is open 90°, the airflow is directed into the interior space horizontally and at the same angle as the exterior airflow.

When the exterior airflow is 45° to the window plane and one sash is open at the same angle and in the same direction, the airflow is directed into the interior space horizontally and at the same angle as the exterior airflow.

Folding Casement Window

The window may be open to a maximum of 100 percent.

The airflow enters the interior space at a horizontal level.

The airflow enters the interior space mainly through the side openings, moderately through the triangular openings at the top and bottom of the window, and slightly between the two sashes.

When both sashes are open to a minimum or maximum, the sashes direct the exterior airflow mainly to each side of the window simultaneously.

Vertical-Pivot Window

The window may be open to a maximum of 100 percent.

The airflow enters the interior space at a horizontal level.

The exterior airflow enters the interior space at the same angle as the sash.

window, unlike the side-hinged casement window, offers an added distinction of being able to take the exterior airflow and spread it wide into the interior space. The vertical-pivot window illustrates the typical characteristic of vertical-vane openings in the extreme case with its air-direction ability. (See Figures 7.29 to 7.50.)

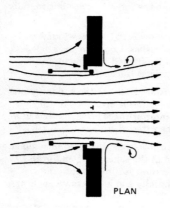

PLAN

Figure 7.29 Both sashes of the side-hinged casement window are completely open. The air flows through the entire area of the window. The interior airstream follows the same direction as the exterior airflow. *(Adapted from Ref. 4)*

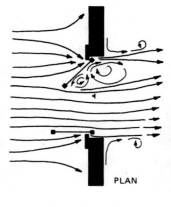

PLAN

Figure 7.30 One sash of the side-hinged casement window is completely open, and the other is open to a 30° angle. The resultant interior airflow is similar to the pattern created when both sashes are completely open. *(Adapted from Ref. 4)*

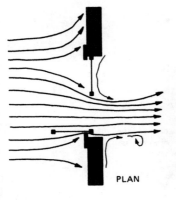

PLAN

Figure 7.31 One sash of the side-hinged casement window is closed, and the other is completely open. The interior airflow is in the same direction as the exterior air movement. This position of the window sashes reduces the amount of air entering the inside space. The results would be identical regardless of which sash were open; only the horizontal position of the airflow would change. *(Adapted from Ref. 4)*

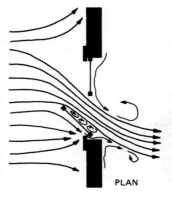

PLAN

Figure 7.32 One sash of the side-hinged casement window is closed, and the other is open at a 30° angle. The positions of the sashes cause the exterior airflow, which is perpendicular to the opening, to be diverted into the same direction as the angle of the sash. *(Adapted from Ref. 4)*

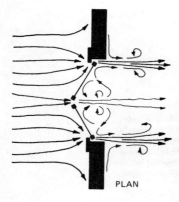

PLAN

Figure 7.33 Both sashes of the side-hinged casement window are open at 30° angles. The main airflow occurs at each side of the window, and minute air movement is present between the two sashes. The overall velocity of the airflow is reduced on the interior side. *(Adapted from Ref. 4)*

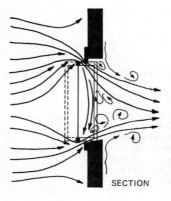

SECTION

Figure 7.34 Both sashes of the side-hinged casement window are open at 30° angles. The resultant air movement pattern is called a vertical "triangular duct" shape. *(Adapted from Ref. 4)*

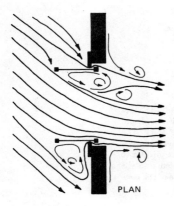

PLAN

Figure 7.35 Both sashes of the side-hinged casement window are completely open. A 45° angle exterior airflow has relatively the same results as a perpendicular airflow. The air flows through the entire area of the window. *(Adapted from Ref. 4)*

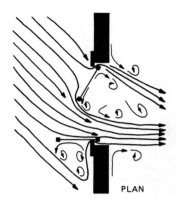

PLAN

Figure 7.36 One sash of the side-hinged casement window is completely open, and the other is open at a 30° angle. The exterior airflow is redirected into a path perpendicular to the window into the interior space. *(Adapted from Ref. 4)*

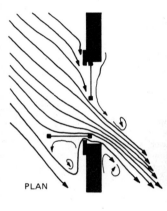

PLAN

Figure 7.37 One sash of the side-hinged casement window is closed, and the other is completely open, which guides the exterior airflow into the interior space. The inside movement of air is in the same direction as the exterior airflow. *(Adapted from Ref. 4)*

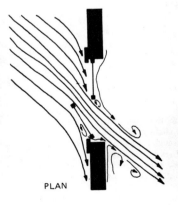

PLAN

Figure 7.38 One sash of the side-hinged casement window is closed and the other is open to a 30° angle, which guides the airflow into the interior space. The results are basically the same whether the sash is open at a 30° angle or completely open. *(Adapted from Ref. 4)*

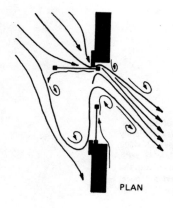

PLAN

Figure 7.39 One sash of the side-hinged casement window is closed and the other is completely open, which blocks the exterior airflow. The position of the sashes reduces the amount of air gaining access into the interior space. *(Adapted from Ref. 4)*

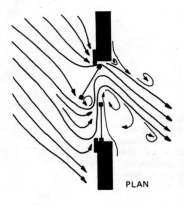

PLAN

Figure 7.40 One sash of the side-hinged casement window is closed and the other is open to a 30° angle, which blocks the exterior airflow. The interior airflow is similar to that when one sash is closed and the other is completely open. *(Adapted from Ref. 4)*

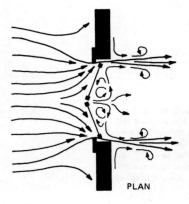

PLAN

Figure 7.41 The folding casement window is open only slightly. Most of the airflow passes through the sides of the window. Regardless of the direction of exterior airflow, the movement of air on the inside will be perpendicular to the window.

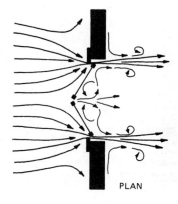

PLAN

Figure 7.42 The folding casement window is open to a 30° angle. Although most of the airflow occurs at the sides of the window, some air turbulence is present in the center of the opening. *(Adapted from Ref. 4)*

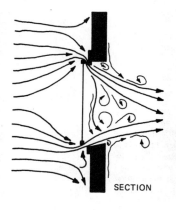

SECTION

Figure 7.43 The folding casement window is open to a 30° angle. The resultant airflow travels in a flue pattern known as a vertical "triangular duct" shape. *(Adapted from Ref. 4)*

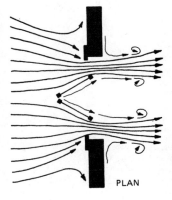

PLAN

Figure 7.44 The folding casement window is fully open. The airflow passes along the sides of the window into the interior space.

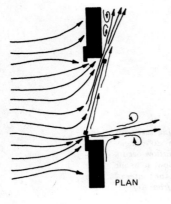

PLAN

Figure 7.45 The vertical-pivot window is open slightly. The angle of the sash creates a severe change in the direction of the airflow at the recessed end of the sash. Air movement continues in relatively the same direction as the exterior airflow at the projected end of the sash.

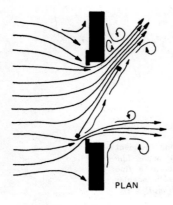

PLAN

Figure 7.46 The vertical-pivot window is open to a 30° angle. The recessed end of the sash causes the exterior airflow to change direction, whereas the projected end does not. The directional change is not as great as when the sash is open only slightly.

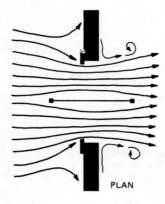

PLAN

Figure 7.47 The vertical-pivot window is fully open. The sash has no significant effect on the perpendicular exterior air movement.

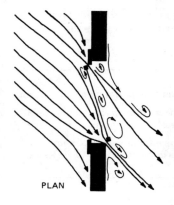

Figure 7.48 The vertical-pivot window is slightly open. The direction of the interior airflow is about the same as that of the exterior airflow. The sash reinforces the angle of the exterior airflow.

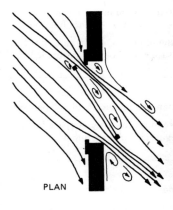

PLAN

Figure 7.49 The vertical-pivot window is open to a 30° angle. The sash permits the exterior airflow to continue to travel in about the same direction.

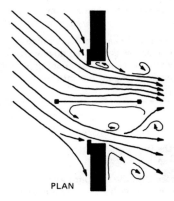

PLAN

Figure 7.50 The vertical-pivot window is fully open. The sash redirects the angled exterior airflow into a perpendicular interior airflow.

Horizontal-vane openings influence the velocity and pattern of air movement, although their versatility is limited (Table 7.3). They have the characteristic of directing air movement upward, which is usually above

TABLE 7.3 Horizontal Vane Openings

Projected Sash Window

> The window may open to a maximum of 100 percent.
>
> The airflow enters the interior space at a horizontal level only when the sash is fully open.
>
> The airflow enters the interior space at an upward direction when the sash is open to any angle less than fully open.

Awning Window

> The window may open to a maximum of 100 percent.
>
> The airflow is directed into the interior space at a horizontal level only when the sashes are fully open.
>
> The airflow enters the interior space in an upward direction when the sashes are open to any angle less than fully open.

Basement Window

> The window may open to a maximum of 100 percent.
>
> The airflow is directed into the interior space in an upward direction regardless of the window's degree of openness.

Hopper Window

> The window may open to a maximum of 100 percent.
>
> The airflow enters into the interior space in a downward direction regardless of the window's degree of openness.
>
> When the window is located near the floor, which is common, the entering airflow is directed downward to the floor. Raising the location of the opening in the wall will direct more of the airflow into the occupant level.

Horizontal-Pivot Window

> The window may open to a maximum of 100 percent.
>
> The airflow enters the interior space at a horizontal level only when the sash is fully open.
>
> The airflow is directed into the interior space in an upward direction when the sash is open to any angle less than fully open.
>
> The airflow travels at the same angle as the sash.

If the bottom of the window projects inward instead of outward, the airflow will be directed to the floor or to the occupant level.

Jalousie Window

The window may open to a maximum of 100 percent.

The sashes, which create laminar air movement, have a stabilizing effect on the airflow as it passes through the opening.

The airflow enters the interior space at the same angle as the sashes.

The airflow is directed into the interior space in an upward direction when the sashes are open to any angle less than horizontal.

The airflow is directed into the interior space at a horizontal level when the sashes are horizontal.

The airflow enters the interior space in a slightly downward direction when the sashes are fully open.

the occupant level. The situation may be rectified by lowering the opening so that the area in need of air movement is above the window. Horizontal air movement is barely obtainable when the window is fully open. The hopper window is the only opening which has characteristics contrary to those of the horizontal-vane type. It directs the airflow downward all the time it is open. The jalousie window is the most versatile of all the horizontal-vane openings. It can guide the airflow upward, horizontally, or downward. (See Figures 7.51 to 7.66.)

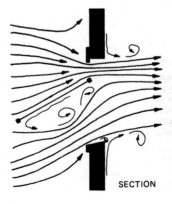

SECTION

Figure 7.51 The projected sash window is fully open. The exterior airflow continues into the interior at a relatively horizontal level. Although the window may be open to a maximum of 100 percent, the area of the airflow on the interior is only about two-thirds of the window's opening. *(Adapted from Ref. 4)*

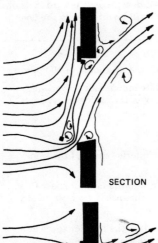

SECTION

Figure 7.52 The projected sash window is slightly open. Any angle less than fully open directs the interior airflow upward at basically the same angle as the sash. *(Adapted from Ref. 4)*

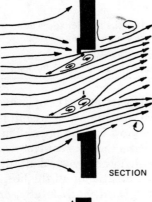

SECTION

Figure 7.53 Both sashes of the awning window are fully open. The interior airflow is relatively horizontal in its pattern. As in the case of the projected sash window, the interior airflow is only about two-thirds as large as the window opening.

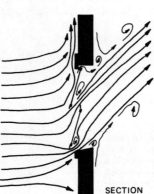

SECTION

Figure 7.54 Both sashes of the awning window are only partly open. Any angle less than fully open guides the resultant airflow upward into the interior space at basically the same angle as that of the sashes.

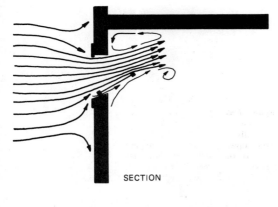

SECTION

Figure 7.55 The basement window has its sash completely open. The interior airflow is directed upward.

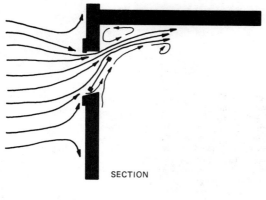

SECTION

Figure 7.56 The sash of the basement window is slightly open. The airflow enters in an upward direction regardless of the angle of the opening. The flow of air follows the same angle as that of the sash.

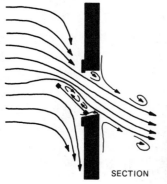

SECTION

Figure 7.57 The sash of the hopper window is fully open. The entering airflow is directed downward regardless of the degree of openness of the sash.

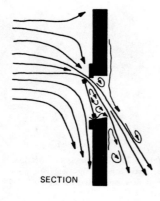

SECTION

Figure 7.58 The sash of the hopper window is only partly open. The interior airflow is directed downward at relatively the same angle as the sash.

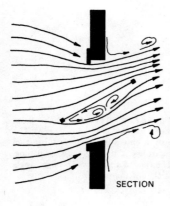

SECTION

Figure 7.59 The sash of the horizontal-pivot window is completely open. The airflow enters the interior space in an upward direction. Even when the sash is fully open, the entering airflow area is about two-thirds as large as the window opening.

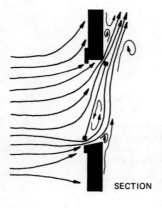

SECTION

Figure 7.60 The sash of the horizontal-pivot window is slightly open. The interior airflow is directed upward at the same angle as the sash.

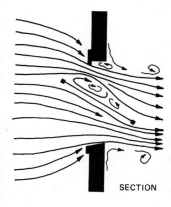

SECTION

Figure 7.61 The horizontal-pivot window is installed in reverse, and the sash is completely open. The entering airflow is directed downward regardless of the degree of openness of the window.

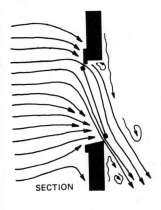

SECTION

Figure 7.62 The horizontal-pivot window is installed in reverse, and the sash is slightly open. The airflow follows the angle of the sash.

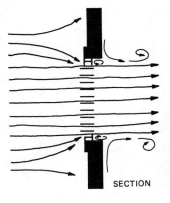

SECTION

Figure 7.63 The sashes of the jalousie window are open to a horizontal position. The airflow follows the angle of the sashes and enters the interior space at a horizontal level. The sashes stabilize the incoming flow of air and establish laminar air movement. *(Adapted from Ref. 4)*

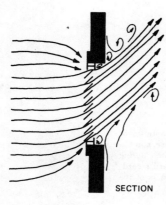

Figure 7.64 The sashes of the jalousie window are open to a 45° angle. The entering airflow is directed upward at relatively the same angle as that of the sashes. *(Adapted from Ref. 4)*

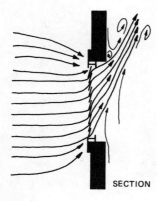

Figure 7.65 The sashes of the jalousie window are slightly open. The airflow enters the interior space in an upward direction at basically the same angle as that of the sashes. *(Adapted from Ref. 4)*

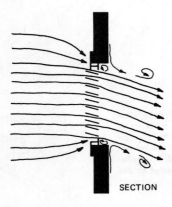

Figure 7.66 The sashes of the jalousie window are fully open. The airflow in the interior space is directed downward. *(Adapted from Ref. 4)*

The large variety of residential designs indicates that there is no set of standards by which window styles can be selected for optimum air movement control. The architectural designer may, however, with a knowledge of how each window style affects air movement, choose the style of window most capable of meeting the specific requirements of the design. The selection would probably include the size, quantity, and placement of windows.

In test studies of conventional windows, the following conclusion was reached: "The air speeds were increased and decreased, and it was found that there was no indication of change in the air patterns." [4] In addition, the air velocity and placement of outlet openings have no apparent effect on the air pattern through the inlet window. Once the airflow enters the interior space in a given direction, it continues in that direction until it encounters an obstruction such as a wall. However, the location of the outlet opening does influence the horizontal flow of air within the room even though it does not affect the vertical air pattern. There is a great opportunity to utilize inlet openings as air movement control devices. They can assist in obtaining optimum air movement in interior spaces through careful selection and placement of each window style by the architectural designer.

Doors

In addition to their main functions doors permit light and air to enter rooms from adjacent spaces. Exterior door designs also include such valuable aspects as good weatherproofing, proper solar heat resistance, structural strength, excellent airtightness, good insulation, easy operation, durable construction and parts, and good air movement control. The valuable aspects of interior door designs are not quite as extensive as those of exterior doors. They involve only structural strength, durable construction and parts, easy operation, and good air movement control.

Whether the door design is for exterior or interior usage is of no real significance in regard to air movement control. The constructional differences do not influence or alter the airflow pattern. Consequently, the only aspect of doors which affects the movement of air is the mode of operation.

Doors for residential use are defined as six basic types; which is used is determined by the manner of opening (Figure 7.67). The *single door* is a one-panel unit which opens by pivoting on a vertical axis along one side (Figures 7.68 to 7.75). It may be either inswinging (opening toward the inside of the space) or outswinging (opening toward the outside of the space). The *double door* is a two-panel unit each panel of which pivots on an individual vertical axis (Figures 7.76 to 7.83). It too may be either inswinging or outswinging. Commercial versions of single and double doors are often capable of swinging in both directions.

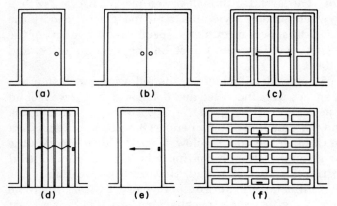

Figure 7.67 Six types of residential doors: (a) single door, (b) double door, (c) bifold door, (d) folding door, (e) sliding door, and (f) garage door.

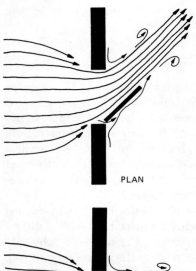

Figure 7.68 The inswinging single door is open to a 45° angle. The airflow enters the interior space at the same angle as the door.

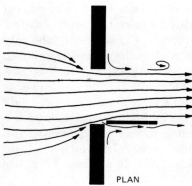

Figure 7.69 The inswinging single door is open to a 90° angle. The entering airflow follows the path of the exterior airflow, which the door does not significantly affect.

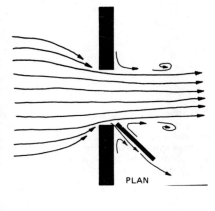

PLAN

Figure 7.70 The inswinging single door is open to a 135° angle, which does not significantly affect the airflow.

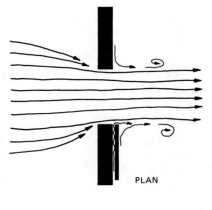

PLAN

Figure 7.71 The inswinging single door is completely open, and the airflow entering the interior space is totally unaffected.

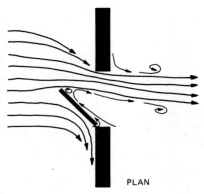

PLAN

Figure 7.72 The outswinging single door is open to a 45° angle. The entering airflow is restricted to about one-third of the total doorway. The door deflects most of the airflow from entering the interior space.

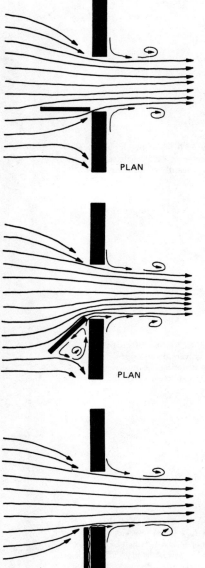

PLAN

Figure 7.73 The outswinging single door is open to a 90° angle. The airflow enters the interior space in the same direction as that of the exterior airflow. The door slightly reduces the total volume of air entering the interior space.

PLAN

Figure 7.74 The outswinging single door is open to a 135° angle. The airflow is funneled into the interior space.

PLAN

Figure 7.75 The outswinging single door is fully open. The airflow entering the interior space is completely unaffected by the door.

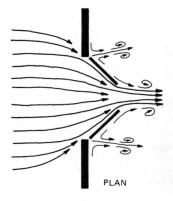

PLAN

Figure 7.76 The inswinging double door is open to a 45° angle. Most of the exterior airflow enters the interior space through the center, and a small amount of air movement gains access at both sides of the doorway. The airflow is funneled into the interior space.

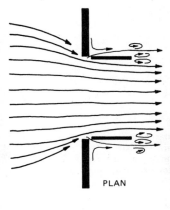

PLAN

Figure 7.77 The inswinging double door is open to a 90° angle. The airflow is directed straight into the interior space. The door has a slight stabilizing effect on the airflow.

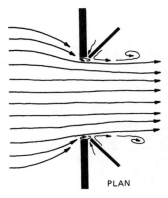

PLAN

Figure 7.78 The inswinging double door is open to a 135° angle. The entering airflow is relatively unaffected by the door.

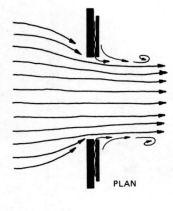

PLAN

Figure 7.79 The inswinging double door is completely open. The airflow entering the interior space is not significantly influenced by the door.

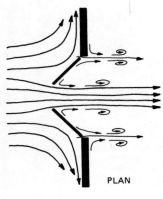

PLAN

Figure 7.80 The outswinging double door is open to a 45° angle. The exterior airflow is directed into three main airstreams, one in the center and one on each side.

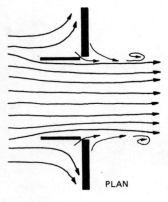

PLAN

Figure 7.81 The outswinging double door is open to a 90° angle. The airflow enters the interior space in the same direction as the original airflow. The door has a slight stabilizing effect on the airflow.

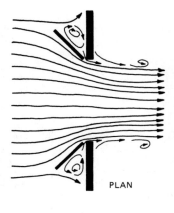

Figure 7.82 The outswinging double door is open to a 135° angle. The exterior airflow is funneled into the interior space.

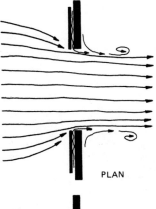

Figure 7.83 The outswinging double door is completely open. The entering airflow is totally unaffected by the door.

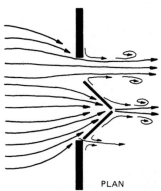

Figure 7.84 The two-panel inswinging bifold door is open to a 45° angle. A portion of the airflow is funneled into a narrow stream, and the remainder enters the interior space directly.

The *bifold door* is a two- or four-panel unit which opens by one panel pivoting on a vertical axis while an attached panel slides and folds open in conjunction with the first panel. The two-panel unit involves one folding system (Figures 7.84 to 7.87), and the four-panel unit involves two

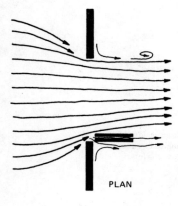

PLAN

Figure 7.85 The two-panel inswinging bifold door is completely open. The airflow entering the interior space is relatively unaffected by the door.

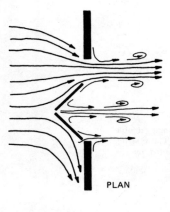

PLAN

Figure 7.86 The two-panel outswinging bifold door is open to a 45° angle. Most of the airflow is directed into the opening; the remainder is directed away from the opening or filters between the two panels.

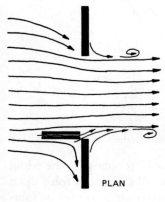

PLAN

Figure 7.87 The two-panel outswinging bifold door is fully open. The entering airflow is little affected because the door blocks off only a small amount of air from traveling through the opening.

folding systems (Figures 7.88 to 7.91). The bifold door may be either inswinging or outswinging. The *folding door* is a multiple-panel unit which opens by unfolding and closes by folding (Figures 7.92 and 7.93). The panels are attached one to another in a sequence. A folding door may be long enough to divide an entire room into two spaces.

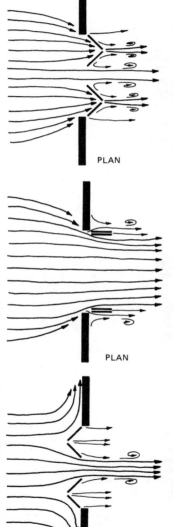

PLAN

Figure 7.88 The four-panel inswinging bifold door is open to a 45° angle. The exterior airflow is divided into three main streams. One stream enters the interior space directly, and the other two are funneled into narrow streams through the panels.

PLAN

Figure 7.89 The four-panel inswinging bifold door is completely open. The airflow enters the interior space in the same direction as the exterior airflow. The door has a slight stabilizing effect on the airflow.

PLAN

Figure 7.90 The four-panel outswinging bifold door is open to a 45° angle. The airflow is directed into a main stream as it enters the interior space.

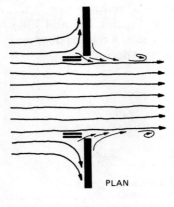

PLAN

Figure 7.91 The four-panel out-swinging bifold door is fully open. The entering airflow follows the direction of the exterior airflow. The door does not significantly affect the airflow, but it does slightly stabilize it.

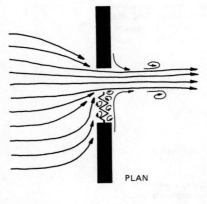

PLAN

Figure 7.92 The folding door is completely open. The airflow is not influenced as it travels through the opening.

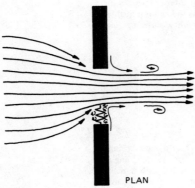

PLAN

Figure 7.93 The folding door is half-way open. The entering airflow is relatively unaffected by the door, which acts as an extension of the existing wall. The door only reduces the amount of air entering the interior space.

The *sliding door* is usually a one-panel unit which opens by sliding in a single plane (Figures 7.94 and 7.95). The door may either pocket inside the wall or slide beside the wall. Two-panel units are used sometimes in residences but are commonly found in commercial establishments. The

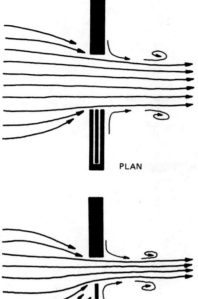

PLAN

Figure 7.94 The sliding door is fully open. The door has no effect on the entering airflow, since it is inside the wall.

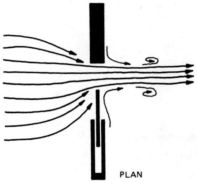

PLAN

Figure 7.95 The sliding door is open midway. The exterior airflow entering the interior space is only reduced in volume, since the door is effectively an extension of the wall.

garage door consists of two main types: sectional and one-piece. The sectional garage door has several panels, attached in a sequence, that open by sliding upward in a vertical plane; at the ceiling, the panels turn and slide horizontally (Figures 7.96 and 7.97). The one-piece garage door is a

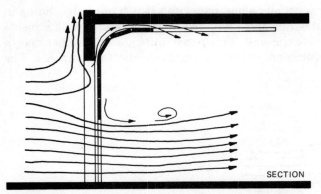

Figure 7.96 The sectional garage door is open halfway. The airflow is directed downward toward the ground level, and the velocity of the airflow is increased.

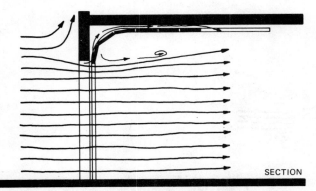

Figure 7.97 The sectional garage door is fully open. The exterior airflow is not affected by the door.

single-panel unit which opens by a dual sliding action of the door (Figures 7.98 and 7.99). The top portion of the door slides horizontally along the ceiling while the bottom portion slides vertically along the door's opening.

Doors for residences are available in a wide selection of types, styles, and modes of operation. One type of door even has numerous positional variations which can alter air movement in a different manner with each new position. Consequently, no one type of door is suited for all situations. The operation of the door, as well as the door type, is critical in maintaining optimum air movement control. The architectural designer may, with an understanding of how each type of door affects air movement, select the type most qualified to fulfill the needs of the design. In addition, the designer may desire to inform the occupants of the residence how the selected type of door benefits them in controlling air movement.

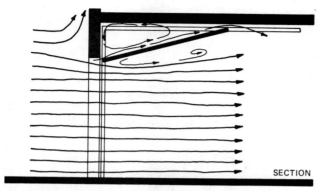

Figure 7.98 The one-piece garage door is open midway. The airflow is divided into two streams: one follows the ground plane, and one follows the ceiling plane.

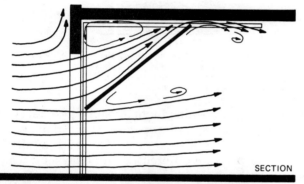

Figure 7.99 The one-piece garage door is completely open. The entering airflow is relatively unaffected by the door.

Roof monitors

The use of skylights, dormers, clerestories, belvederes, cupolas, and other roof monitors as air movement control devices has been relatively insignificant (Figure 7.100). In most northern climates, where heating is an issue, roof monitors are added to residential designs as methods of gaining extra heat; in southern climates such as Florida, they are used to lessen the heat load on a house. The native Haitians employ a central dormer to ventilate their houses. Actually, the use of roof monitors to expel excessive heat from residences is extremely logical. When air is heated, it rises and collects under the ceilings of most houses. An opening in the roof or upper portions of the walls permits the heated air to escape and allows cool air to replace it.

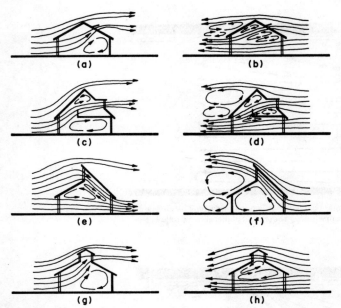

Figure 7.100 Openings in the roof may provide a release for heated air trapped under the ceiling through either direct air movement or secondary air currents. Skylights, (a) and (b), dormers, (c) and (d), clerestories, (e) and (f), and belvederes, (g) and (h), are types of openings in the roof.

A fan may also be added to the opening to increase the heat-releasing effect. Roof monitors could function as outlet openings while windows and doors at the occupant level operate as inlet openings. The resulting airflow would combine the forces of pressure differential and buoyancy, which may be more effective in moving air than simple cross-ventilation. During the design phase of residential buildings, architectural designers should consider the potential benefits of roof monitors.

Vents and ventilators

Attic spaces are common in most residential structures. They receive the direct impact of solar heat, which may be transferred down into the living spaces. Consequently, attic ventilation is one of the most important aspects of structural cooling. In addition to releasing heat from the attic space, attic ventilation prevents condensation during the winter. The heated air in the living spaces, which contains moisture, penetrates the insulated ceiling and comes in contact with the cold air of the attic space. The water vapor condenses onto the roof structure and insulation, resulting in moisture damage. Therefore, openings in the attic space should allow adequate air movement throughout the space to prevent the buildup of heat and condensation.

By itself, movement of air within an attic space is not sufficient for the removal of heat and moisture; elimination of the problems involves the amount of air moving through the space and the uniformity of the air movement. A good rule of thumb is to provide 1 sq ft of vent opening for each 300 sq ft of attic floor area to obtain minimum air movement. The airflow should occur throughout the entire attic space to prevent dead air pockets where heat and moisture might collect. Fulfilling the two requirements necessitates proper placement of adequately open vents.

Several ventilation techniques are generally utilized in residential structures, but they are not equally effective in obtaining thorough air movement. Eight basic techniques are employed either alone or in combination. They include gable louvers, soffit vents, ridge vents, stationary ventilators, rotary ventilators, mechanical ventilators, and whole-house fans.

Gable louvers are usually triangular vents located at both gable ends of a house. They are the most popular devices in use because they are inconspicuous and inexpensive. When exterior air movement occurs in alignment with gable louvers, the airflow travels through the attic in a narrow stream. When exterior air movement is perpendicular to gable louvers, minute air movement is created at each vent, and no cross-ventilation occurs (Figure 7.101). Under optimum conditions, gable louvers are capable of lowering the attic temperature as much as 13 percent, which is poor in comparison with other techniques.

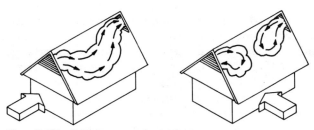

Figure 7.101 Gable louvers produce airflow patterns that differ with the direction of the exterior airflow. [5]

Soffit vents located in the eaves bring air movement only along the attic floor. When the exterior airflow is parallel to the soffit vents, air travels across the attic floor at the same width as the vents. When the exterior airflow is perpendicular to the soffit vents, the movement of air occurs mainly at each venting system. The airflow enters and exits along the same soffit vent (Figure 7.102). The attic temperature, measured at the floor, may be lowered as much as 18°F, or 13 percent, when optimum air movement is achieved in the attic space.

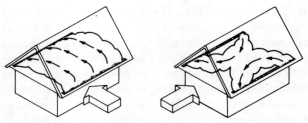

Figure 7.102 Soffit vents establish various air movement patterns along the attic floor in conjunction with the exterior airflow. [5]

A combination of gable louvers and soffit vents is still an inadequate method of introducing air movement to the attic space. Both sets of vents tend to act independently of each other (Figure 7.103).

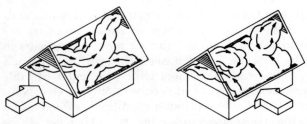

Figure 7.103 Gable louvers and soffit vents produce several different patterns of air movement; however, the combination does not produce adequate attic ventilation. [5]

Ridge vents used by themselves provide no air movement to the attic space. When they are used in conjunction with soffit vents, however, an excellent quantity and uniformity of airflow throughout the entire attic space is secured (Figure 7.104). Regardless of the exterior airflow direction with respect to the ventilation system, the attic receives air movement in every part. In order to achieve optimum performance, one should make the net free area of the ridge vents equal to the net free area of the soffit vents.

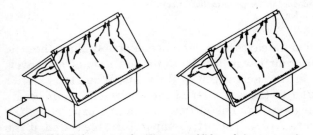

Figure 7.104 Ridge vents and soffit vents establish good air movement in every portion of the attic space. [5]

The system of continuous **ridge and soffit vents** induces a convective flow of air even when no outside air movement is detectable. Hot air rises up and out through the ridge vents. That creates negative pressure near the peak of the roof, which draws cooler air into the attic through the soffit vents (Figure 7.105) . The system reduces the temperature of the attic space as much as 35°F on a still air day, and its effectiveness remains relatively constant until the exterior air reaches a speed of 10 mph.

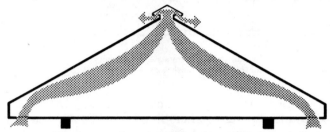

Figure 7.105 The ridge-and-soffit vent system utilizes both buoyancy and pressure differential forces to drive air movement through an attic space.

The ridge-and-soffit vent system appears to function more efficiently when conditions for obtaining human comfort become worse. Although the efficiency of the system drops as the exterior air movement increases, the attic space temperature reduces further, since the increased exterior airflow drives more air through the attic (Figure 7.106).

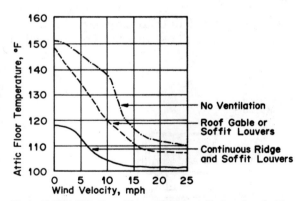

Figure 7.106 Effective attic ventilation results in reduced attic space temperatures. [5]

Stationary ventilators include such devices as gravity ventilators, low-pitch slant roof ventilators, roof mushroom ventilators, and stack vents (Figure 7.107). These are nonmechanical nonmoving devices that allow heat to escape from the attic space. Some are often used in conjunction with fans for more effective heat removal from either the attic or the living spaces. Some lists of stationary ventilators include pipes. Tests performed by a group of engineers revealed that an "uncapped pipe" provides better attic ventilation than a roof turbine, and the investigators also concluded that "a simple air shield [a type of cap] having a 125° arc produced the highest performance index" of stationary and rotary ventilators. [6]

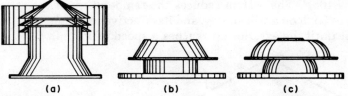

(a)   **(b)**   **(c)**

Figure 7.107 Stationary ventilators are manufactured in many styles. The gravity ventilator (*a*), low-pitch slant roof ventilator (*b*), and roof mushroom ventilator (*c*) are the most common types.

Rotary ventilators are similar to nonmechanical stationary ventilators except that they usually move in a circular motion (Figure 7.108). The roof turbine is the most common type used for residential buildings. The revolving ventilator and free-flow ventilator are other popular types, but they are usually found on commercial establishments. The revolving action of these ventilators is created by the passing of exterior airflow over the roof; in other words, rotary ventilators are wind-powered. Once the devices are set in motion, they develop a negative pressure area within the attic space in the immediate vicinity of the ventilator. Consequently, cool air is drawn into the attic through other vents, such as soffit vents, and hot air is pulled out by the rotary ventilator. Although rotary ventilators are more effective in the removal of heated air than stationary ventilators when the exterior airflow exceeds 10 mph, both function equally well in still air situations.

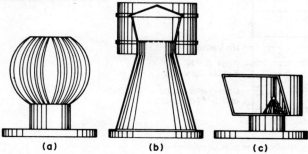

(a)   **(b)**   **(c)**

Figure 7.108 Rotary ventilators include the roof turbine (*a*), free-flow ventilator (*b*), and revolving ventilator (*c*).

The *mechanical ventilator* most frequently used is a fan; attic fans and power ventilators are the two most common types (Figure 7.109). The attic fan is a combination of a large fan and louvers, usually located in the gable end of a house. A power ventilator is a stationary ventilator with a fan added. In either case, the energy required to power the fan usually offsets the savings gained through a cooler attic. Passive ventilation systems, or nonmechanical techniques, provide significant savings as a return for the cost of establishing the system.

(a) **(b)**

Figure 7.109 Mechanical ventilators are popular methods of reducing the heat load in residential attics. They include the attic fan (*a*) and the power ventilator (*b*).

Whole-house fans are not typical attic ventilators. They move air through both living and nonliving spaces (Figure 7.110). A whole-house fan creates a mechanical stack effect: It draws cool air into the living spaces through the windows while the warmer air is emptied out through an upper space such as the attic. It is capable of taking advantage of the cool night air by bringing it into the house to expel the building's radiant heat. In a very short time, the whole-house fan may pay for itself by maintaining human comfort without resort to air conditioning. In fact, a properly sized whole-house fan uses only about one-fifth the energy a typical air conditioner requires.

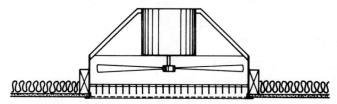

Figure 7.110 A whole-house fan is usually mounted in the ceiling of a residence.

Using attic ventilation techniques obviously provides human comfort by lowering the temperature of attic spaces. However, it would appear probable that a thoroughly ventilated attic would place a greater energy load on the heating equipment and reduce the residence's effectiveness to maintain comfort during the cold months. Winter conditions actually leave the attic space only a few degrees warmer than the outside temperature regardless of the lack of or the amount of attic venting. In addition, adequately insulated ceilings permit no significant heat losses from the living spaces. Therefore, architectural designers should plan for summer since the benefits of attic ventilation are much more pronounced in hot weather conditions.

Specially designed openings

Extra openings are sometimes provided in residences to reduce the heat load on the structure. Most of them depend on buoyancy as the means of moving the air. Although it is weaker than pressure differentials and may prove insufficient to provide human comfort because of its low air velocity, buoyancy may benefit the occupants indirectly by diminishing the effect of the sun's rays on the house. Thermal chimneys and double-skin constructions utilize buoyancy forces to move air even though pressure differentials may affect the airflow.

Thermal chimneys are becoming more popular in residences as a natural alternative to a whole-house fan (Figure 7.111.) In simple terms, a thermal chimney is a box that sits on the roof or a chimney attached to a sunny wall. The top extends beyond any other high points of the roof, and located within it is a louvered vent. The shaft is open to the interior spaces through operable openings. The chimney contains glass along the south side in the upper portion, so the sunlight can enter and heat the air. The heated air rises up and out the chimney, which pulls cool air into the house, if a window is open, and into the chimney. Thermal chimneys function best during the hottest part of a day; at night they do not work at all. Structural cooling is the most important aspect, since thermal chimneys do not provide enough air movement to supply human comfort.

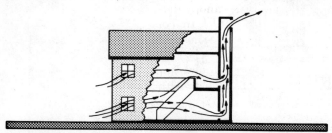

Figure 7.111 Thermal chimneys supply air movement within structures through temperature differences or buoyancy.

Double-skin construction has been utilized in southern residences for many years. The walls and/or roof are built with an interior wall or roof, and a lightweight skin is attached to the outside (Figure 7.112). The sunlight hits the outer skin and is converted to heat which is transferred to the air space, usually 1 to 3 in, between the skin and wall. The air in the space becomes heated and rises up through the cavity. Cooler air enters vents at the base of the double-skin wall or roof while the heated air exits through a vent at the top.

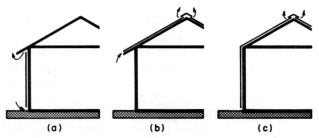

Figure 7.112 Double-skin construction may involve a wall only (a), a roof only (b), or a combination wall and roof (c).

The system greatly reduces the amount of heat reaching the interior spaces. Experiments of double-skin houses in Florida indicate an increased efficiency of wall insulation value by about 30 percent over conventional homes. Even in cold weather, the system functions well. The secret to the success of double-skin construction is to locate the insulation on the exterior of the inside walls. If concrete block is used for the interior walls, it will store coolness in the summer and heat in the winter. Consequently, the interior spaces may remain relatively constant in temperature.

Roof canopies are a version of double-skin construction, except that two roofs are built instead of one (Figure 7.113). A separate structure usually supports each roof. The higher roof or canopy blocks the sun's rays from reaching the second or actual roof of the building. The canopy may capture and channel exterior air movement between the roofs. Daylight may enter the building through skylights or transparent surfaces in both roofs. The temperature beneath the canopy itself may be as much as 20°F lower than that of the outside air.[7] In comparison to double-skin construction, roof canopies provide the added benefit of sheltered outdoor spaces.

Specially designed openings furnish new approaches for an old problem: the avoidance and removal of heat. Although the techniques are

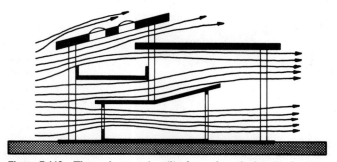

Figure 7.113 The roof canopy is utilized to reduce the heat load on the structures below.

subtle in their approach to providing air movement, they are extremely effective on hot, still days.

Openings in buildings establish distinct micro-climatic environments within the interior spaces. They permit sunlight and air to enter and exit structures. Those two elements affect the occupants of the buildings as they alter the buildings' micro-climates. Sunlight reduces the requirement of electric lighting. Carefully planned and placed openings can provide the correct types of sunlight, direct or indirect, and the amount in areas needing light. Air movement is necessary for the health and wellbeing of the occupants. Air quality and human comfort are primarily determined by the proper guidance of air movement. Furthermore, the energy consumption of buildings is lessened by appropriate air movement control. Openings provide planned visual releases for the occupants, which may enhance the character of the interior spaces. In short, building openings are effective devices which improve the quality of life without devouring vast quantities of energy.

REFERENCES

1. B. Givoni, *Man, Climate, and Architecture,* Applied Science, Ltd., London, 1976.
2. William W. Caudill and Bob H. Reed, *Geometry of Classrooms as Related to Natural Lighting and Natural Ventilation,* Research Report No. 36, Texas Engineering Experiment Station, College Station, Tex., 1952.
3. "Window Design Report Gives Energy Strategies," *AIA Journal,* vol. 66, September 1977, p. 92.
4. Theo R. Holleman, *Air Flow through Conventional Window Openings,* Research Report No. 33, Texas Engineering Experiment Station, College Station, Tex., 1951.
5. Burt Hill and Associates, *Planning and Building the Minimum Energy Dwelling,* Craftsman Book, Solana Beach, Calif., 1977.
6. Subrato Chandra et al., *Passive Cooling by Natural Ventilation: A Review and Research Plan,* Florida Solar Energy Center, Cape Canaveral, Fla., 1981.
7. David Morton, "The Elements and Form," *Progressive Architecture,* vol. 62, April 1981.

8

Opening Modifications

Features outside the building and near inlet openings influence the airflow pattern to a large extent. Although the openings themselves modify the entering airflow, external elements further alter the velocity and pattern. Horizontal projections, vertical extensions, screens and louvers, and vegetation affect the airflow prior to or simultaneously with the inlet opening. The modifications may be in a variety of patterns and velocities. Opening modifications, if properly used, may greatly enhance the movement of air or hinder it.

Horizontal Projections

roof overhangs

Horizontal projections are parallel to the ground plane and located near inlet openings. They include porches, overhangs, awnings, sunshades, and canopies. In addition to providing shade, horizontal projections establish some unique air movement patterns. Their extension beyond the wall surface can help capture the exterior air movement and bring it into the interior spaces.

William W. Caudill and Bob H. Reed performed some original test studies involving horizontal projections during the 1950s (Figure 8.1). Further test studies have added minute information. The original findings indicate that horizontal projections may drastically alter the pattern of the airflow, especially overhangs placed directly over the inlet openings. An inlet opening with a horizontal projection acts as a nozzle directing the flow of air. In addition, the projection may increase the amount and velocity of air entering the interior space. The overall size and shape of a projection determine how much air flows into a building through the opening.

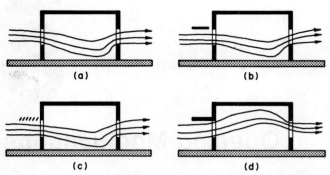

Figure 8.1 Different types of overhangs immediately over the inlet opening create various air movement patterns. The simple opening with (*a*) no overhang, (*b*) slotted overhang and (*c*) louvered overhang provide a downward airflow, and (*d*) a solid overhang establishes an upward airflow. *(Adapted from, Ref. 1)*

Although horizontal projections are usually located above inlet openings, they also influence the air patterns of openings placed above them (Figure 8.2). The projection slope and depth, fascia size, and opening height are the factors which determine the potential of air movement entering a window located above a horizontal projection. In actual architectural practice, the airflow is usually directed upward to cool the structure rather than downward to provide evaporative cooling for the occupants.

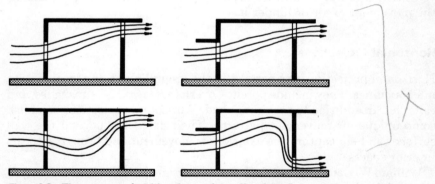

Figure 8.2 The presence and position of an overhang affect the airflow patterns in the interior space. Note that the outlet opening has no influence on the pattern of the airflow in the vertical plane. *(Adapted from Ref. 1)*

Vertical Extensions

Vertical extensions are valuable devices for air movement control as well as being popular shading elements. They usually protrude from the side of the building and affect both the velocity and pattern of the airflow (Figure 8.3). They are most efficient when they project out as far as the opening is wide. Wing-walls are typical vertical extensions which are utilized to establish positive pressure areas near inlet openings.

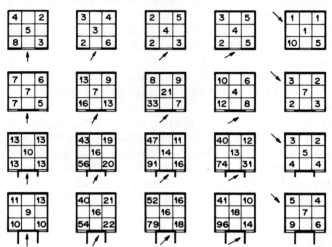

Figure 8.3 Interior air velocities in buildings with vertical extensions of various depths compared to those in buildings without vertical extensions. The numbers represent the relative air speed as percents of the exterior air velocity. [2]

In some situations, two vertical extensions may be employed, as when wing-walls are added to a building with two openings facing the exterior airflow. One wall creates positive pressure at one opening while the second creates negative pressure at the other opening. Consequently, the exterior airflow enters the positive pressure opening (the inlet) and exits the negative pressure opening (the outlet). If the exterior airflow is slightly skewed to the openings, the influence of the wing-walls becomes more effective.

Shutters may also be used as vertical extension devices. Traditionally, they are flush with the building's exterior walls, but there is a trend toward repositioning them. When they project from the wall, they create shade for the windows and direct air movement into the building. Such shutters are analogous to side-hinged casement windows: They both reach out and guide the exterior air movement into the building's interior.

Vertical extensions provide a unique approach to the design of residential structures. The designer can develop a climatically efficient building and locate the areas to receive air movement. He or she may determine the placement of the openings and modify them to guide the airflow into desired paths.

Screens and Louvers

Screens and louvers affect both airflow pattern and velocity. They are usually placed in planes parallel to the windows. Insect screens are the most common type; homeowners expect to find them installed in all window openings. In addition to serving their primary purpose, they reduce the incoming airflow. Consequently, screened areas receive less air movement and may experience increased air temperatures. Insect screens may reduce the initial airflow velocity by 25 to 50 percent when it is between 2 to 10 mph. [4] However, screening an area in front of an inlet opening produces better air movement through the opening than screening the opening itself.

Sun screens have basically the same effect on air movement as insect screens. Louvers and venetian blinds exercise control over both the pattern and velocity of the airflow. They have the capability of guiding the flow of air in almost any direction through the adjustment of their slats. In that respect, they are basically the same as jalousie windows.

Vegetation

Hedges, shrubs, and trees affect the micro-climatic environments within and outside buildings. They establish windbreaks, filter dust, prevent erosion, reduce noise, and provide shade—they improve air quality. They also have beneficial psychological effects on people by creating a sensation of freshness and coolness. More important, vegetation controls the movement of air, which provides a direct perception of human comfort.

The micro-climatic environment of a building is altered by vegetation located near building openings. Plantings may either increase or reduce the velocity of the airflow through a structure. The direction of the airflow within the building may also be modified. Vegetation situated near the structure's outlet openings has no effect on the movement of air through the building unless the outlet openings are partly or completely obstructed.

The American Society of Landscape Architects Foundation funded a book which illustrates how vegetation, given specific conditions, affects air movement within buildings. Vegetation alters the movement of air by filtering, redirecting, obstructing, and deflecting the initial airflow. These sudden changes caused by the vegetation develop new positive and negative pressure areas which, in turn, produce new airflow patterns. (See Figures 8.4 to 8.17.)

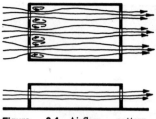

Figure 8.4 Airflow pattern within a building without vegetational influence. [4]

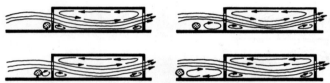

Figure 8.5 A low hedge, less than 3 ft high, is introduced at various distances from the inlet opening. [4]

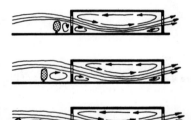

Figure 8.6 A hedge about 5 ft high positioned at several distances from the inlet opening. [4]

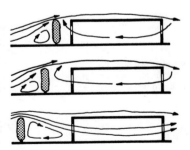

Figure 8.7 A hedge with height equal to that of the building and located at various distances from the inlet opening.[4]

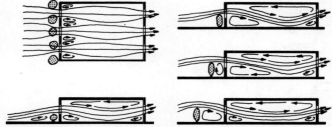

Figure 8.8 Low shrubs, 3 ft high, and medium shrubs, 5 ft high, positioned at various distances from the inlet openings. [4]

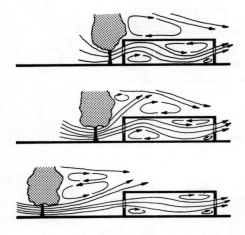

Figure 8.9 A tree located at several distances from the inlet opening. [4]

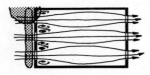

Figure 8.10 A low hedge, less than 3 ft high, located next to the inlet opening and a tree positioned 5 ft from the building. [4]

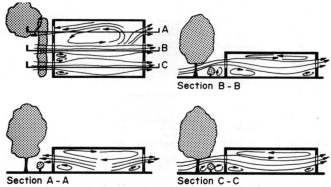

Section B - B

Section A - A   Section C - C

Figure 8.11   A low hedge, less than 3 ft high, located 5 ft from the inlet opening and a tree located 10 ft from the building. [4]

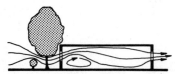

Figure 8.12   A low hedge, less than 3 ft high, 10 ft from the building and a tree 5 ft from the building's inlet opening.[4]

Figure 8.13   A low hedge, less than 3 ft high, 30 ft from the inlet opening and a tree 20 ft from the building. [4]

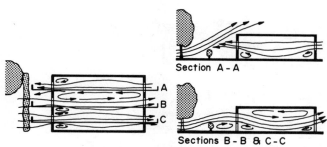

Section A - A

Sections B - B & C - C

Figure 8.14   A low hedge, less than 3 ft high, located 10 ft from the inlet openings and a tree 20 ft from the building's corner. [4]

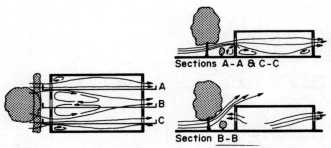

Figure 8.15 A low hedge, less than 3 ft high, 5 ft from the inlet openings and a tree 10 ft from the building. [4]

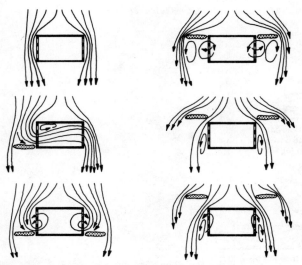

Figure 8.16 Medium and high hedges located at various positions in relation to the building. [4]

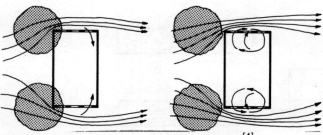

Figure 8.17 Trees at various locations near a building. [4]

Modifications of the building's openings establish alternative methods of air movement control. The modifications, however, affect only the inlet openings and not the outlet openings. In short, opening modifications may fine-tune or significantly alter the velocity and pattern of air movement entering architectural structures.

REFERENCES

1. William W. Caudill and Bob H. Reed, *Geometry of Classrooms as Related to Natural Lighting and Natural Ventilation,* Research Report No. 36, Texas Engineering Experiment Station, College Station, Tex., 1952.
2. William W. Caudill, Sherman E. Crites and Elmer G. Smith, *Some General Considerations in the Natural Ventilation of Buildings, Research Report No. 22,* Texas Engineering Experiment Station, College Station, Tex., 1951.
3. B. Givoni, *Man, Climate, and Architecture,* Applied Science, Ltd., London, 1976.
4. Subrato Chandra et al., *Passive Cooling by Natural Ventilation: A Review and Research Plan,* Florida Solar Energy Center, Cape Canaveral, Fla., 1981.
5. Gary O. Robinette and Charles McClennon, *Landscape Planning for Energy Conservation,* Van Nostrand Reinhold, New York, 1983.

9

Interior Modifications

Air movement in interior spaces of a building is governed by several physical elements. The inlet openings direct the initial pattern of the airflow; the outlet openings regulate the airflow velocity. The dimensions and divisions of the spaces modify the pattern and velocity of air movement through the reorganization of pressure differentials and the alteration of airflow inertia. The latter two elements determine the major airflow inside a building once air movement gains access to the interior.

Interior spaces are contained within the overall scale of a building. Whenever the building depth is greater than the depth of its spaces, the spaces require conjunctive interaction to obtain thorough air movement. They may be connected directly or through transitional spaces such as hallways. The configuration and arrangement of the spaces may vary tremendously. As the exterior airflow enters the house, it is guided through the house in numerous paths and directions. The changes reduce the initial velocity of the airflow while distributing the movement of air throughout the entire house. The impact of interior space dimensions and divisions may either reduce and distribute or amplify and densify the air movement.

The geometry of a building form plays a significant role in the movement of air through the building's interior as well as around and over its exterior. Once the structure's exterior form and openings introduce the airflow into the interior spaces, the dimensions and divisions of the spaces redirect the pattern and modify the velocity of the air.

Space Dimensions

The configuration of the building's interior spaces may effectively assist in the control of air movement within those spaces. The dimensions of the spaces regulate the pattern and velocity of the airflow in predictable ways having a high degree of accuracy even though slight variations may occur because of minor obstructions such as furniture. The length, depth, height, and shape of the interior spaces are the factors in air movement control.

The length of the interior space has varying effects on air movement. When the inlet opening is the whole length of the space and the exterior airflow is perpendicular to the inlet opening, the length of the space influences air movement only minutely (Figure 9.1). Even when the exterior airflow is skewed to the inlet opening, the airflow is basically unaffected by the length of the space. However, the area of calm in the interior space is proportionately smaller relative to the size of the space. This eddy, like most calm areas, has little effect on human comfort.

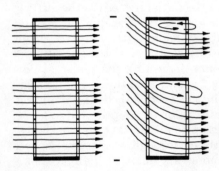

Figure 9.1 Whether the exterior air flow is perpendicular or askew to the opening, the length of the interior space has little effect on air movement when the inlet openings are the full length of the space.

On the other hand, when the inlet opening is only a small proportion of the space length and the exterior airflow is skewed to it, the length of the space significantly influences air movement (Figure 9.2). When the exterior air movement is perpendicular to the inlet openings, the length of the space has no important effect on the air movement. Increasing the length of an interior

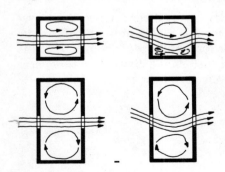

Figure 9.2 When the inlet opening is only a small portion of the building and the exterior airflow is perpendicular to the inlet, the length of the interior space has little effect on air movement. When the inlet opening is a small portion of the building and the exterior airflow is askew to the inlet opening, the length of the interior space has a significant influence on the airflow.

space generally adds no meaningful benefits to that space with regard to air movement.

The depth of an interior space is unlikely to have any significant influence on the movement of air through the space. The airflow travels through more interior space, but that has very little effect on either the pattern or the speed (Figure 9.3). The only difference is the number of air changes. In a space with greater depth than another space air replacement takes longer, but the velocity or pattern of the airflow is about the same in both spaces. "It is airflow, not air change that counts." [1] Regardless of the number of air changes in a space, the occupants experience no comfort if there is no air movement.

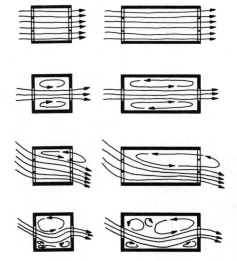

Figure 9.3 The depth of the interior space has little influence on the pattern and velocity of the airflow. However, a space that is twice the depth of another space doubles the volume of air and has half the number of air changes. It is the movement of air through a space, nonetheless, which helps to create human comfort.

The height of an interior space does not significantly affect airflow but there is a greater potential for buoyancy to occur as the height is increased (Figures 9.4 and 9.5). The gain in height allows for an amplification of thermal differentials, and that facilitates the vertical movement of air. Variances in pressure forces usually move air in a horizontal plane, and those airflows are not influenced by differences in the interior space height. "Tests conducted on an 8-foot ceiling situation gave identical results as those conducted on a 12-foot ceiling ... In each case, the inlet openings were proportionately located, and the air flow was similar." [1] Since the height of an interior space does not significantly affect horizontal airflow but does influence vertical air movement, greater interior height would probably profit occupants only on still hot days.

In regard to air movement within interior spaces, there appears to be a contradiction. As stated earlier, the geometry of a building plays an

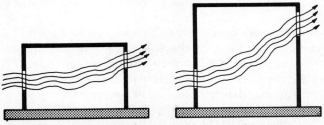

Figure 9.4 An increase in vertical height within an interior space permits the buoyancy forces to influence air movement to a greater extent.

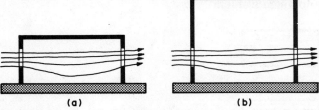

(a)   (b)

Figure 9.5 A space with a low ceiling (*a*) and a space with a high ceiling (*b*) produce similar airflow patterns.

important role in the movement of air through the interior space (Figure 9.6). However, the length, depth, and height of the space affect airflow only in an insignificant manner. Here the key element is geometry. The shape of the interior space sides, top, and bottom influences the air movement within the space. It is the changing of length, depth, and height which controls and guides air movement inside a building (Figures 9.7 and 9.8).

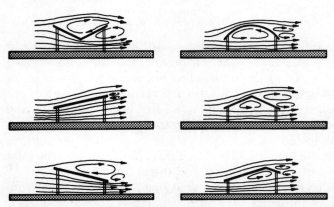

Figure 9.6 The height of a space influences both the pattern and velocity of an airflow changing in form and relation to the ground plane.[2]

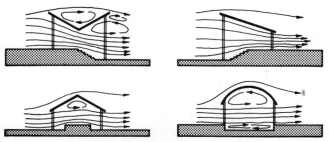

Figure 9.7   The ceiling and ground planes may function in unison in controlling the airflow pattern and velocity.

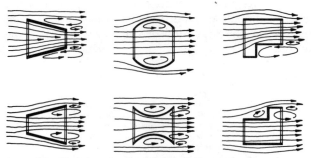

Figure 9.8   Changes in the shape of the wall planes alter the air movement pattern and velocity.

The control of air movement within a space is not determined by any single factor alone but is derived from the interaction of factors. The flow of air is a dynamic volumetric function; consequently, it must also be guided and directed in the same three-dimensional sphere. The planes of an interior space such as walls which determine the length, depth, and height of the space are relatively two-dimensional. The thickness of these planes is basically insignificant with regard to air movement control. Therefore, the control of air movement is achieved from the volumetric relationship of these planes—the shape of the space.

Space Divisions

Once air movement is introduced into an interior space, its velocity and pattern are influenced by space divisions. Although partitions affect air movement patterns to a significant degree, their greatest impact is on airflow velocity. Consequently, partitions must be so placed and arranged that airflow through the space is optimized (Figure 9.9).

Figure 9.9 Partitions within an interior space may work with or against the airflow.

The proper location of space divisions may encourage more effective air movement. And although partitions interrupt the flow of air, they may also be used to redirect or guide airflow (Figure 9.10). An important factor to remember is that any change in the direction of airflow reduces the velocity and internal energy of airflow. In addition, the greater the initial velocity of airflow, the greater the reduction of velocity when the flow encounters an obstruction.

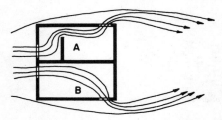

Figure 9.10 Partitions which cause abrupt changes in the direction of the airflow reduce the initial velocity of the movement of air. A partition located near the inlet opening (a) slows the speed of the airflow in comparison with an inlet opening without a partition (b). [1]

B. Givoni investigated the effects of subdividing interior spaces into two unequal spaces (Figure 9.11). In all partition arrangements and opening placements the airflow direction was perpendicular to the inlet opening and velocities of the interior spaces ranged from 5 to 98 percent of the

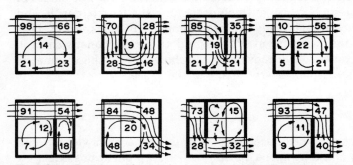

Figure 9.11 Partitions within interior spaces have various effects on air movement patterns and velocities as the exterior airflow enters the inlet opening perpendicularly. The numbers represent the relative air speeds as percents of the exterior air velocity. [3]

exterior airflow. "The velocities were the lowest when the partition was in front and nearer to the inlet window, as the air had to change direction upon entering, but better conditions were obtained when the partition was nearer the outlet." [4] According to the test results, a partition in an interior space should be located nearer the outlet opening for optimum air movement through a space. Also, thorough air movement occurs throughout an entire space when the inlet opening and the outlet opening are not in direct alignment with regard to the exterior airflow.

Space divisions may occur in horizontal and/or vertical planes. Although vertical plane partitions, such as walls, are more commonly dealt with in building forms, horizontal-plane partitions, such as kitchen cabinets, are equally effective in controlling air movement (Figure 9.12). Even household furniture may create friction and eddies in the flow of air. Consequently, by proper placement and arrangement of space divisions, furniture, and other potential obstructions, optimum air movement throughout the space may be achieved. Although maximum airflow has been sacrificed, it is not completely lost.

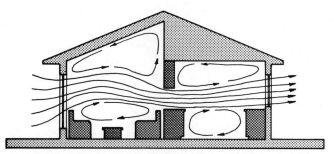

Figure 9.12 Partitions may be in the horizontal as well as the vertical plane.

The velocity and pattern of airflow are directed by the dimensions, shapes, and divisions of the interior spaces. The velocity may be increased or decreased, and the pattern may remain in its initial direction or its course may be altered. Nevertheless, the air must be allowed to enter an interior space in the first place. Consequently, the principal controlling device of air movement within the space is the inlet opening.

REFERENCES

1. William W. Caudill and Bob H. Reed, *Geometry of Classrooms as Related to Natural Lighting and Natural Ventilation*, Research Report No. 36, Texas Engineering Experiment Station, College Station, Tex., 1952.
2. William G. Wagner et al., *Shelter for Physical Education*, Texas Engineering Experiment Station, College Station, Tex., 1961.

3. William W. Caudill, Sherman E. Crites, and Elmer Smith, *Some General Considerations in the Natural Ventilation of Buildings, Research Report No. 22,* Texas Engineering Experiment Station, College Station, Tex., 1951.

4. B. Givoni, *Man, Climate, and Architecture,* Applied Science, Ltd., London, 1976.

10

Appropriate Technology

Air movement within interior spaces occurs by natural means, induced techniques, or mechanical means. When air is no longer set into motion by natural or induced means, mechanical means may be utilized. At this point, technology is introduced.

In this century, technology has progressed by leaps and bounds. The advances provide a great many conveniences, many of which are more sophisticated than their users. In general, the devices require highly trained specialists for their repair and maintenance. The costs of operating and maintaining equipment are increasing.

Appropriate technology is the suitable utilization of mechanisms. First, the technology must fit the required work and the task it replaces. For example, no one would replace an exhaust fan with a 40,000-Btu air conditioner to ventilate a typical residential bathroom.

Low technology, intermediate technology, and high technology are the categories commonly used to identify the level of technology in regard to energy. Economy of scale occurs when the proper technology is matched to an equal-energy-level task. Second, the serviceability of devices is important; the devices should be operable by their owners and require minimum maintenance. If servicing must be performed, local skill should be available. Third, but not last, the operation of the equipment must not bankrupt the owner. Similarly, maintenance should be only a minor inconvenience rather than a catastrophe. Finally, appropriate technological devices should pay for themselves in a reasonable period of time related to the life cycle of the device. If they do, technology has been applied appropriately and is a convenience and not a burden.

Appropriate air movement control technology ranges from high to low energy levels depending on the task to be performed. The application of devices becomes necessary when natural and induced forces no longer provide the air movement necessary to create comfort. The energy level of the technology usually increases with the degree of discomfort. Therefore, appropriate air movement technology varies from simple fans to elaborate air conditioners.

Fans

Electric fans function on relatively low levels of energy. Low-technology or passive elements such as a double-skin roof usually require no outside energy to function. In a typical house, a fan will normally keep the occupants of the home comfortable up to temperatures in the middle 80s (Figure 10.1).

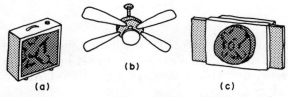

(a)   (b)   (c)

Figure 10.1   One of the three most common electric fans may be found in most homes: (*a*) portable fan, (*b*) ceiling fan, and (*c*) window fan.

Portable, or box-type, fans are efficient and economical. The portable fan may be used to "flush out" the house by facing it outward in a window. The fan creates negative pressure around the window opening which induces airflow in open windows located in opposite or adjacent walls. The same effect is created with a window fan, which is made to fit inside a window opening.

Ceiling fans, or paddle fans, are variable-speed, wide-blade fans mounted in the ceiling (Figures 10.2 and 10.3). They usually stir the air into motion within the entire room, and they have the added benefit of circulating heated air in the winter. During the winter, when windows are shut, the heated air which rises to the ceiling can be circulated back down into the occupant level. During the summer, when the windows are open, new air can be drawn into the space by the ceiling fan.

Attic fans and whole-house fans represent two totally different strategies. Attic fans ventilate only the attic, whereas whole-house fans ventilate the entire structure (Figure 10.4). Attic fans reduce attic heat gain which would transfer to the lower living spaces. Without an attic fan or a passive ventilation system, the attic temperature may be 40°F above the outside ambient air temperature. To illustrate, on a typical summer day in Florida, the outside temperature in the shade may easily reach 98°F;

consequently, an unvented attic could have a temperature of 138°F. Burt Hill and Associates demonstrated that the temperature of a typical unventilated attic may reach as high as 150°F at ceiling level.

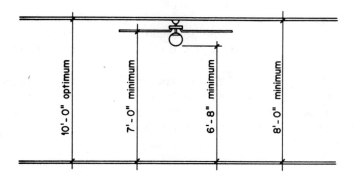

Figure 10.2   Minimum clearances for a ceiling fan are necessary to assure safe and proper operation of the unit and obtain successful results. The ideal ceiling height is 10 ft for optimum benefit of air movement created by the ceiling fan. *(Adapted from Ref. 1)*

Figure 10.3   The size of the ceiling fan, from blade tip to blade tip, determines the volume of air which may be cooled effectively. A 36-in ceiling fan can cool up to 200 sq ft or 2000 cu ft, and a 52-in ceiling fan can cool up to 400 sq ft or 4000 cu ft. *(Adapted from Ref. 1)*

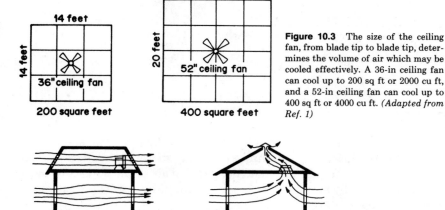

Figure 10.4   (a) Attic fans ventilate only the attic; (b) whole-house fans ventilate the complete house.

Attic fans, which are as effective as a continuous ridge and soffit vent system, do not directly affect comfort through air movement. Whole-house fans affect both the living and non-living spaces. They are a mechanical version of the stack effect because they draw air from the lower spaces to the upper spaces. Cool air is brought into the house through the windows, and the warmer air is emptied out through the attic space. The whole-house fan can take advantage of cool night air to exhaust the radiant heat of the structure.

As a champion energy saver, the whole-house fan uses only about one-fifth as much energy as an air conditioner if it is correctly sized (Figure 10.5). It should have a cubic feet per minute rating of one-half to one and one-half times the volume of the living spaces. [2] The greater the cubic feet per minute rating within the two limits the better the air movement. Fans may also be utilized in conjunction with other systems, and the effectiveness of the combined systems depends on their ability to function as a unit.

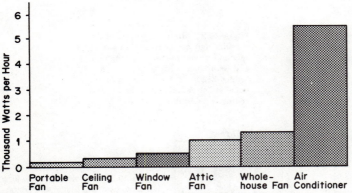

Figure 10.5 Fans offer a tremendous advantage over air conditioning in saving energy and money while cooling a house.

Cool Tubes

Cool tubes are a relatively new idea still in the experimental stage. They are long pipes buried in the ground which draw in warm outside air, transfer the heat to the earth, and then deliver cool air to the house. Actually, the system is really more complex in design, and the results are extremely varied (Figure 10.6).

The degree to which the air from the cool tubes is chilled depends partly on the local soil conditions and partly on the cool tube design. The nature of the soil determines the cool tube ability to take heat from the warm

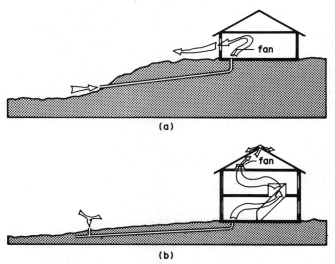

Figure 10.6 The installation of cool tubes may involve several pipes or a single pipe. Either specific blower fans at the cool tube's exhaust (a) or a whole-house fan with windows closed (b) may operate the cool tubes.

air as the air travels through the tube. In addition to soil conditions, the design of the system, such as the material, diameter, length, depth, location, and proximity of the pipes, influences the effectiveness of the cool tubes.

Fans are used to draw cool air out of the tubes, since natural forces are insufficient. Care should be exercised here since an improperly sized fan could cost more to run than a full-size air conditioner. If the system functions continuously, it will exhaust the earth's ability to absorb heat, usually within an hour. Some cool tubes in actual use are often used to give a quick blast of cooling which can eliminate a large portion of heat retained in a residence.

Most existing installations of cool tubes have been designed by seat-of-the-pants technicians. The problems usually encountered are high humidity, limited cooling ability, odor-producing mold and mildew, radon infiltration, and cost versus benefits.

Evaporative Cooling

Evaporative cooling is a technique of utilizing the evaporation of water, which absorbs heat from the air. Three different methods are employed.

In the first method, called a volume cooler, movement of air over a body of water such as a pond, pool, or fountain is used (Figure 10.7). The airflow may be natural, induced, or mechanical. As the warm air mass travels over the water surface, the air is cooled through the evaporation of the water. The rate of cooling is greater for an active volume of water, such as a stream or fountain, than for a still body of water.

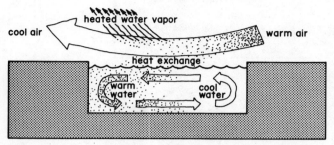

Figure 10.7 The volume cooler method involves the flow of warm air over a body of water, which extracts heat from the air and produces water vapor. The rate of heat exchange depends on the thermal differential of the air and water and on the velocity of the airflow.

The Crystal Cathedral, in Garden Grove, California, utilizes the volume cooler system. The building functions as a massive thermal chimney which uses surrounding pools and fountains to cool the entering air. The cathedral's mechanical engineer, Marvin Mass, "never dreamed it would work as well as it did." [3] An energy analysis, funded by the United States Department of Energy, revealed that the inside temperature will fall within the comfort range of 68 to 78°F at a probability of 41 percent for the morning services and 42 percent for the evening services; when the comfort range is extended to 60 to 85°F, the probability changes to 88 percent for the morning services and 83 percent for the evening services. [4] According to the analysis, the cathedral is more likely to be too cold than too hot (Figure 10.8).

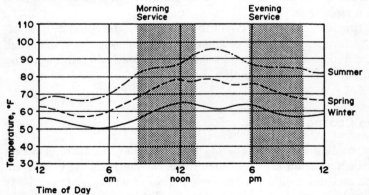

Figure 10.8 The Crystal Cathedral volume cooler system generally provides comfort for the occupants throughout the year as shown by the energy analysis results. *(Adapted from Ref. 4)*

The second system, an evaporative cooler, is an intermediate technological device in comparison to the volume cooler. A fan, pump, and
fibrous pad are the components. The system works as follows: a fibrous
pad, usually made of cellulose or plastic fiber, is stood up vertically and
a small pump trickles water over it while a blower draws air through it
(Figure 10.9). Cool, damp air is created as the evaporating water lowers
the ambient temperature of the air passing through the pad. Evaporative coolers work well in most climates and especially well in arid ones.
However, in a hot humid climate, the resultant air may feel too damp
for comfort.

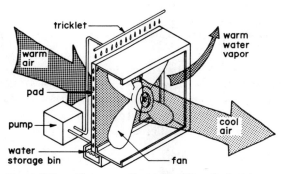

Figure 10.9   The evaporative cooler utilizes the heat-trap-
ping characteristics of water to remove heat from a flowing
stream of air.

The third alternative, the two-stage evaporative cooler, is really two
separate systems used together (Figure 10.10). One system works the
night shift, and the other system works 24 hr a day. The night shift system
functions identically to the evaporative cooler except that it is limited to
nighttime. The 24-hr system draws cool night air directly through an
evaporative cooler and into a rock storage bin which stores the coolness.
In the morning, the night shift unit is shut off, and the other system
switches its mode of operation. During the day, air is brought out of the
cooled-rock bin and is passed through an evaporative cooler as it enters
the living spaces. Consequently, the air is cooled twice. Although the two-
stage evaporative cooler is expensive to construct, its low operational cost
may offer a payback in only a few years in a reasonably dry climate. It
might be worthwhile to experiment with a rock storage bin without an
evaporative cooler as an intermediate method.

Absorption Chillers

Absorption chillers are not a new saving concept. They are a type of air
conditioner which is solar-powered rather than driven by an electric

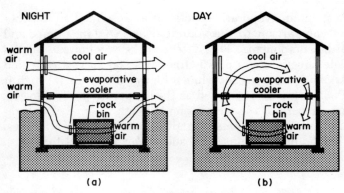

Figure 10.10 The two-stage evaporative cooler utilizes two independent systems. Both units are used during the night (*a*), when one cools the house and the other charges up for the following day (*b*).

compressor. This high-technology device, based on the principles of gas refrigeration of the 1930s and 1940s, is driven by thermal differentials created by solar heat. Absorption chillers, at present, are custom-made and are not likely to be competitive in the marketplace because of the recent advances in desiccant-evaporative chillers.

Desiccant-Evaporative Chillers

The newest technology in solar cooling is desiccant-evaporative chillers commonly known as solar air conditioners. Although it sounds like a contradiction, the hotter the exterior air becomes, the better they work.

The device first removes moisture from the outside air to create hot, dry air and then evaporates moisture back into it to produce cool, moist air (Figure 10.11). The key element is the ability to utilize the latent heat of vaporizing water. The design is simple and moving parts are few, which increases its potential to succeed in the marketplace even though it requires solar collectors and a large water storage tank.

The desiccant-evaporative chiller has the added benefit of always supplying fresh air and removing stale air; it thereby decreases the probability of indoor air pollution. Today, several prototypes are being tested in anticipation of marketing in the late 1980s.

Thermal Exchangers

Thermal exchangers are commonly known as air-to-air heat exchangers even though the systems can be utilized to heat or cool interior spaces (Figure 10.12). They are especially used in airtight houses, where contaminants are trapped because of a low infiltration rate. The exchanger introduces new, fresh air into the house without the heat losses or gains incurred by opening windows.

Air Intake   Exhaust

95°	hot moist air	OUTSIDE	hot moist air	120°
	moisture vapor lossed	DESICCANT WHEEL	moisture vapor added	
			hot dry air	180°
155°	hot dry air	SOLAR REGENERATION COIL	heat added	
			hot dry air	150°
	heat lossed	HEAT EXCHANGER WHEEL	heat added	
78°	warm dry air		warm dry air	75°
	moisture vapor added	EVAPORATIVE ELEMENTS	moisture vapor lossed	
68°	cool moist air	INSIDE	warm moist air	78°

Supply Air   Return Air

Figure 10.11 The desiccant–evaporative chiller functions as a sophisticated thermal exchanger with great cooling capacity.

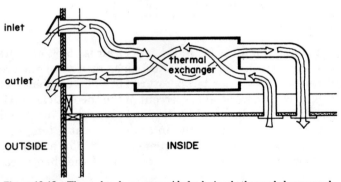

inlet

outlet

OUTSIDE   INSIDE

Figure 10.12 Thermal exchangers provide fresh air, whether cooled or warmed, and lessen the demand for huge mechanical systems.

The thermal exchange system functions as follows: Outside air is drawn into the system through a duct or vent, and a fan pulls the airflow through the exchanger along thermal transfer surfaces. At the same time, a blower drives the interior air through the exchanger on the opposite side of the thermal-transfer surfaces (Figure 10.13). The result is that incoming air either absorbs heat from the outgoing air or the exiting air extracts heat from the entering air. Consequently, the thermal exchanger may cool or heat the air entering the house.

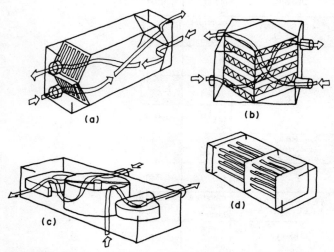

Figure 10.13 The core of a thermal exchanger is made in a variety of ways. In the counterflow model (*a*), the two streams of airflow in opposite directions and parallel to each other. The two air streams in the crossflow unit (*b*) flow parallel in a diagonal manner. In the rotary model (*c*), a slow-moving revolving wheel with air compartments provides the means of thermal exchange. The heat pipe method (*d*) utilizes a refrigerant inside the pipes, which transfers the heat to one end of the pipe and coolness to the other end while two separate air streams travel over different ends of the pipes. Regardless of which type of core is employed, the results and efficiencies are relatively equal. (*Adapted from Ref. 5*)

For example, if the outdoor air is 30°F, the indoor air is 75°F, and the thermal exchanger supplies air to the interior space at a temperature of 55°F, the unit raises the exterior temperature 25°. The efficiency of the thermal exchanger is then rated at 55 percent of the theoretical maximum, 45°. Similarly, if the outdoor air is 95°F, the indoor air is 70°F, and the thermal exchanger supplies air to the interior space at a temperature of 80°F, the unit lowers the exterior temperature 15°. The efficiency of the thermal exchanger is then rated at 60 percent of the theoretical maximum, 25°.

As the airflow rate increases, the efficiency of the unit decreases. Filters may be utilized along with the system as a means of controlling pollutants in the air or adding moisture to and subtracting it from the air.

Thermal exchangers vary from small window units to large equipment that requires ducts. Likewise, retail prices for the devices range from $200 to $1500, with installation and ducts extra. However, manufacturers claim that the units will pay for themselves as heat exchangers in 2 to 5 years, since device efficiency ranges from 70 to 90 percent. Lawrence Berkeley Laboratory, on the other hand, tested 10 models and obtained efficiencies from 45 to 85 percent, which would mean a payback period of 5 to more than 30 years. [5]

The real value of a thermal exchanger is twofold: The amount of energy saved by utilizing such a unit may reduce the amount of fuel consumed for heating and cooling, and the severity of pollution within the house environment may be lessened if filters are employed with the system. Consequently, thermal exchangers, in the long run, may add dollars to homeowner pockets and years to lives.

Solar Exchangers

Solar exchangers, which include Trombe walls and air solar panels, utilize the sun's energy to gain heat for interior spaces or the night sky radiation to expel heat from the building (Figure 10.14). They are basically a box, space, or room with one side glazed. Smaller units, such as air solar panels, usually contain fins, baffles, or flat plates that function as heat-transfer surfaces. Large units, such as Trombe walls, consist of brick, stone, water drums, or some other form of heat-transfer mass. Typically, a system utilized for heat-gaining purposes can obtain a temperature of 100°F or better on a cold sunny day. Regardless of how solar exchangers are installed, they are relatively equal in their efficiencies.

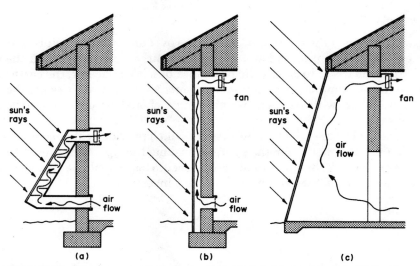

Figure 10.14   The numerous solar exchanger designs include (a) air solar panels, (b) Trombe walls, and (c) greenhouses.

Air Conditioners

When all else fails, an air conditioner may be used. It is one of the easiest but most expensive ways to cool interior spaces. It consists of a fan and refrigerant which extracts coolness from the hot outside air and places it within the house. If an air conditioner must be used, the operation of small units, such as window air conditioners, may be considered for individual spaces, since usually only one space at a time is occupied.

One version of the air conditioner is the heat pump, which has the ability to cool or heat interior spaces. An earth-coupled heat pump is better than a standard air-to-air heat pump (Figure 10.15). The system extracts heat or coolness from water pumped from the ground and returns

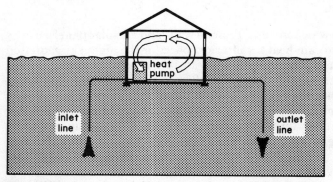

Figure 10.15   An earth-coupled heat pump is the most efficient of all air-conditioning units.

the wastewater to the ground. Its advantage over an air-to-air unit is that the earth's temperature is lower and significantly more stable than the air's. In comparison to an electric resistance baseboard heater, the earth-coupled heat pump is 3 times more efficient. As far as priorities go, the air conditioner, with its high use of energy, should be considered the last resort.

Mechanical forces used to drive systems which provide heat or coolness should deliver benefits for the energy consumed in operating them. In other words, the technology applied to a specific task should be appropriate to the comfort derived and the amount invested.

REFERENCES

1. "Replace a Light with a Fan," *Southern Living,* July 1982.
2. Frederic S. Langa, "Many Ways to Cut It," *New Shelter,* July/August 1982.
3. Robert E. Fischer, "The Crystal Cathedral—Embodiment of Light and Nature," *Architectural Record,* vol. 168, November 1981.
4. Vladimir Bazjanac, "Energy Analysis: Crystal Cathedral, Garden Grove, California," *Progressive Architecture,* vol. 61, December 1980.
5. V. Elaine Smay, "Heat-Saving Vents—Are They the Solution to Indoor Pollution?," *Popular Science,* January 1983, and Eugene Thompson, illustrator.

11

Techniques of Air Movement Control

Architectural designers with an understanding of air movement control have an opportunity to create climatically responsive residential structures. The utilization of air movement principles and techniques may contribute to relatively comfortable homes year-round. In addition, the air quality within residences may be greatly improved while the building's energy consumption is significantly reduced. In short, good air movement control can improve the quality of life in residential structures.

Architecture structures are based on a variety of fundamental issues. The priorities are different for every design problem as seen by the designer attempting to resolve the concerns at hand. Tables 11.1 and 11.2 summarize the principles and techniques of air movement control.

TABLE 11.1   Primary Principles of Air Movement

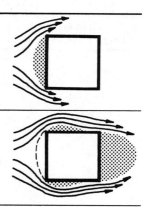

Positive Pressure Areas: An obstacle, such as a house, in the path of air movement will obstruct the airflow and cause it to pile up and slow down until it locates a new path to follow. The affected area is designated as a positive, or high, pressure area. The walls adjacent to the positive pressure area should contain the inlet opening that allows the airflow to be forced into the building by the pressure exerted on the wall.

Negative Pressure Areas: As the airflow completely surrounds the building, negative pressure areas, or wind shadows, are created. The size and shape of the area are determined by the configuration and scale of the building. The walls adjacent to the negative pressure areas should contain the outlet openings that permit the airflow to be drawn out of the house by the negative pressure areas.

Inertia: Air will flow through an opening if another opening is provided and travel within the building is in the same direction as the exterior air movement until the interior airflow encounters an obstruction.

Pressure Differences: The difference between the positive pressure area and the negative pressure area helps determine the airflow velocity through the building as well as the exterior airflow velocity, inlet and outlet opening sizes, and directional changes in the airflow.

Directional Changes: A change in the direction of the airflow consumes energy from the airflow and reduces its velocity. The greater the velocity of the airflow when the change of direction occurs, the larger is the loss of energy and velocity.

Optimum Airflow: Inlet openings and outlet openings should be as large as possible in order to optimize airflow. When the air movement is perpendicular as it encounters the inlet and outlet openings in alignment, the airflow will pass through the building in a narrow stream *(a)*. The remainder of the interior space will receive no significant air movement. The result will be maximum airflow in a minimum area. Skewed air movement will result in optimum airflow in a maximum area *(b)*.

(a)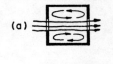

(b)

Maximum Velocity: The maximum velocity of air movement is obtained when the outlet opening is larger than the inlet opening. In that situation, positive pressure is created on the windward face of the building and negative pressure is established inside the building; consequently, increased air movement occurs within the interior spaces *(a)*. If the exterior airflow is reversed, positive pressure is created inside the building and negative pressure is established on the leeward face of the building; consequently, increased air movement occurs outside the building *(b)*.

Opening Locations: The airflow pattern within a building is determined by the placement of the inlet opening as well as by the initial airflow direction and the location of positive and negative pressure areas.

TABLE 11.2 Air Movement Control Techniques

Character of Air Movement

Air movement is a volumetric function, or three-dimensional in character.

Pressure, an action of force differentials, may cause air movement.

Buoyancy, a temperature and density differential, may establish air movement.

Buoyancy forces are weaker than pressure differential forces.

Buoyancy and pressure differentials may function together or separately.

When buoyancy and pressure differentials work together, their resulting force is the square root of the sum of the two forces.

Air movement occurs in a positive-to-negative fashion whether the movement is due to buoyancy and/or pressure differentials.

Air movement is stronger in higher latitudes.

The velocity of air movement increases with height.

Air turbulence may increase with height at the microclimatic scale.

The velocity of air movement is usually lower in summer than in winter.

 Prevailing winter air movement usually comes from a different direction than prevailing summer air movement.

Wind is air movement due to the forces of nature.

Ventilation is air movement through a building.

The depth of a calm area is independent of the air movement's velocity.

Comfort

Human comfort is determined by seven variables: velocity of the air movement, temperature of the air movement, relative humidity, surrounding thermal conditions, duration of exposure, degree of exposure or level of clothing, and physical and psychological state of the individual.

Effective cooling is produced by four methods: lower ambient air temperature by direct cooling or heat removal, reduction of the air's moisture content, introduction of air movement, and evaporation of water without the addition of heat.

The cooling effect is proportional to the square root of air movement.

Greater effective cooling results from air movement in high-humidity areas when a temperature is moderately high.

Air movement reduces the heat load on the human body through increased convection and evaporation.

Air movement, at low temperatures, will increase discomfort by removing additional heat from the body.

Air Quality

Natural ventilation introduces fresh air into interior spaces.

Ventilation disperses smoke, dust, and odors.

The residential site should be located in an unpolluted area or windward of a source of pollution.

Energy

Orientation affects the energy consumption of a building.

Building configuration, including length, width, and height, affects energy consumption.

Air movement increases heat loss from a building, which may be desirable or undesirable.

The introduction of outside air into a building, when its temperature is lower than that of the inside air, is a method of cooling.

Shading reduces the heat load on a building.

The use of natural lighting reduces the internal heat load on a building.

Areas with high humidity should not add moisture to large volumes of air as a means of cooling the air mass.

Site

The site should offer opportunities for energy conservation.

The site should have a year-round temperature close to or lower than the temperatures desired, if possible.

The site should allow the best building orientation and configuration for maximum air movement.

Air movement in urban areas is 20 to 30 percent less than in rural areas.

The site should have good air quality to enhance natural ventilation.

Vegetation may be used to enhance air movement and to lower air temperature.

Topography may be used to enhance air movement.

Ponds, lakes, and fountains may be used to reduce the air temperature.

Topography

Topography features should enhance air movement.

Landforms may direct the pattern of air movement.

Air movement on the crest of a hill may be 20 percent greater than on flat land.

Fluctuations in the movement of air increase in intensity as the ground becomes rougher.

The rougher the terrain, the greater the vortex on the windward side of a building.

Cold air settles into low areas and valleys at night.

Earth berms may direct air movement over buildings.

Vegetation

A selected site should have vegetation that enhances air movement.

Vegetation type, location, and placement should be suited for controlling air movement in all directions.

Vegetation may deflect and increase the velocity of air movement.

Vegetation establishes microclimates by altering air temperature, humidity, and movement.

Deciduous trees provide summer shade and air movement control.

Evergreen trees provide shade and air movement control year-round.

An effective windbreak provides a calm area for a distance approximately eight times its height.

The best windbreak is a belt of moderately dense and mixed-species trees.

A windbreak of trees may reduce the velocity of air movement by as much as 50 percent.

A well-planted hedge is better than a fence or wall; a hedge allows a gentle air stream to filter through the foliage.

Trees planted near a building should be kept a distance from the building equal to the building height.

Vegetation reduces heat loads on buildings and surrounding areas.

Vegetation should be located between buildings and paved areas to reduce reflected heat.

Adjacent Structures

Adjacent structures should enhance the positive effects of air movement.

If a building is more than double the height of its neighbors and is six stories or more in height, air movement around the building may become hazardous to persons on the ground.

Fences

The shape and configuration of fences do not significantly affect air movement.

The average porosity of fences determines the character of the downstream airflow.

The height of fences determines the height of the eddy immediately behind the fence.

A solid fence creates a strong pressure differential and high turbulence.

Orientations of Buildings

The location of a building should induce air movement into the building.

The location of a building should minimize adverse air movement.

The location of a building should minimize energy consumption.

Air movement around a building may be two to three times greater than the free-flow air velocity.

Buildings should not be oriented for a particular air movement direction; they should be designed for effective air movement in prevailing directions.

Building orientation influences the location of inlet and outlet openings.

Air pressure is positive on the windward side of the building, which results in an inflow of air into the building.

Air pressure is negative on the leeward side of the building, which results in an outflow of air from the building.

The highest air movement around a building occurs in the windward vortex and the corners.

Shapes of Buildings

The geometry of a building determines the size and shape of the eddy.

Building shape may enhance air movement into and around the building.

Small changes in building form may create large changes in air movement.

Building configuration may assist in cross-ventilation.

Air movement may not reach deep into a building's space.

Building height, width, and depth determine the size of the eddy, or calm area.

Pressures on a roof depend on shape and height of the roof.

A 20° pitch roof, in most cases, is unaffected by pressure created by air movement.

Building shape may increase heat loss.

Exotic shapes may be considered for their energy-saving characteristics.

Buildings elevated on columns or with overhanging upper floors have increased heat loss.

Tall buildings are less likely to be shaded and protected by surrounding vegetation and buildings.

A building's exterior surface area may be increased for greater heat loss.

Textures, vines, fins, wing-walls, and recesses in the building's form may assist in air movement control.

Openings in Buildings

If air movement is desired in a room, inlet and outlet openings should be present.

Air movement may range from relatively small to extremely large amounts, depending on sizes of openings.

Air can be brought into a building at almost any desired point, carried through the living spaces, and exited through other designated points by careful placement of openings.

Openings determine the placement, arrangement, velocity, and pattern of the airflow.

Arrangement, location, and control of openings should combine, rather than oppose, pressure differential and buoyancy forces.

For effective ventilation within a building, openings should be oriented in several directions.

Openings should be located in more than one wall if possible.

Openings should be oriented to take advantage of at least the prevailing air movement.

Cross-ventilation should be encouraged.

Negative pressure at the leeward side of a building draws air out of the building.

Openings in the vicinity of the neutral pressure zones are the least effective for ventilation.

The velocity of air movement entering inlet openings will always be less than the exterior air velocity but seldom less than one-half of the exterior air velocity if adequate outlet openings are provided.

A 25 percent drop in the velocity of the airflow may occur when the inlet opening is located below the average windowsill height of 36 in.

Openings in opposite walls at a high level may create air movement at that level without producing any appreciable air movement at the lower level.

Large floor-to-ceiling openings are ideal for maximum ventilation.

When air pressures around an inlet opening are symmetrical, the airstream enters the interior space in a straight path.

When inlet openings are smaller than outlet openings, increased air velocity may occur near the inlet opening within the space.

An increase in the size of inlet and outlet openings creates an increase in the average air velocity within the space.

Openings should be capable of directing the airflow to where it is desired.

The air movement pattern within a room is determined by the placement of the inlet and outlet openings as well as the initial wind direction and the location of high and low pressures.

The location of inlet openings may greatly affect the pattern of the air movement within interior spaces.

When the inlet opening is larger than the outlet, a dam effect is created within the interior space, which slows down the air speed while maximum air speed occurs just outside the outlet opening.

The wall adjacent to the high-pressure area should contain the inlet openings through which the air will be forced because of the pressure exerted on the wall.

Outlet openings may be located in the leeward side of a building, in the negative pressure areas of the sidewalls.

Outlet opening size, shape, and location have little influence on room ventilation patterns, since they are in the wake of the airflow.

Outlet openings usually control the velocity of the airflow.

Inlet openings usually control the direction of the airflow.

Inlet or outlet openings should not be obstructed.

Openings should be accessible to and operable by occupants.

Air movement in interior spaces should benefit the occupants.

Operable windows may be used to control airflow through the building.

The velocity of the airflow may be varied by operable windows in the outlet openings.

Windows should be designed to permit the greatest possible air movement when needed for cooling.

Operable windows should be used in climates where the exterior air conditions are close to the desired indoor temperature.

Dormers may function as ventilators.

A ridge-and-soffit vent system may prevent dead air from pocketing in the attic where moisture or heat might collect.

Attic ventilation may reduce summer heat buildup and winter moisture condensation in the attic.

Opening Modification

Overhangs may deflect more air into a room.

The slope, depth, and size of overhangs, as well as window dimensions, determine the amount of air that will flow through a clerestory window above the overhang.

Overhangs are capable of completely changing the manner in which the air will flow through a simple opening.

Fixed or adjustable projections may guide air movement into poorly ventilated spaces.

Wing-walls added to windows in only one side of a room may increase the average room's air speed up to 35 percent of the exterior air speed.

Louvers can direct moving air where it will have the greatest cooling affect.

Insect and sun screens may reduce the overall ventilation rate by 25 percent to 50 percent for airspeeds in the 2- to 10-mph range.

Interior Spaces

Air movement in interior spaces is determined by the velocity of the exterior airflow, the angle of the exterior airflow direction, and the area of the openings.

Airflow through a room depends on pressure gradients, orientation of openings, size and placement of windows, inertia of air masses, interior partitions, and types of windows.

The air pattern in a room depends mainly on the initial direction of the exterior airflow.

An open plan may improve air movement through several spaces.

The length of a space has very little effect on the inside airflow when the outside airflow is perpendicular to the room.

The length of a room creates a larger eddy in proportion to the room if the outside airflow is diagonal to the room.

The depth of a room has very little effect on air movement.

The depth of a room does effect the number of air changes for a given period of time; however, air changes are unimportant with regard to comfort.

Ceiling heights, low or high, do not significantly affect airflow through a space.

The shape of the ceiling may affect the airflow pattern depending on its exact profile.

The higher the ceiling, the greater will be the possibility of naturally induced thermal air currents.

Breezeways may act as funnels and accelerate the velocity of the airflow.

Halls should receive natural light and air movement.

Partitions, furniture, and equipment may either aid or hinder natural ventilation.

Interior doors may have louvers to permit air movement between spaces.

12

Techniques Applied
to Actual Building Designs

Architectural designers are constantly influencing the environment by creating new spaces, both interior and exterior. Regardless of the degree of sophistication involved in the creation of the new living areas, climate interacts with structures. Of all the climatic elements, the movement of air has the greatest potential for altering the existing microclimate of a region. Architectural designs can affect the environments of particular sites whether they are located directly on the sites or on adjacent properties.

Designers develop buildings on the basis of particular architectural issues. Their concerns and priorities vary with each design situation as they set about to resolve specific design problems and achieve desired architectural goals. The variety of issues involved in design solutions may range from a selected few to a large number. The delicate balance of these issues is often known only by the designer. However, the issues of energy, climate, topography, vegetation, mechanical systems, structural systems, economy, operation, function, sociality, psychology, and physiology are usually impacted by the movement of air.

A comparison of the theory and practice of current architectural design with respect to air movement control reveals an innate incompatibility. Five diverse residential designs developed by architectural designers have been selected for study as to their potential for positive air movement control, the capability of the building to guide and redirect the airflow patterns to benefit the occupants. The selection of the designs was based on the individual characteristics of the design solution as discussed later and the fact that the house was designed by an experienced architect.

Each design is analyzed, evaluated, and discussed with regard to actual airflow patterns in the original design and in the modified design used for experimental purposes.

Each residential design was submitted to smoke chamber tests, and the resultant airflow patterns were analyzed. The angles of attack—the angles of the direction of airflow with respect to the plane of the building side facing the airflow—utilized in these studies were 90° and 45° (Table 12.1). Although actual air movement rarely approaches a building at a true 90° or 45°angle, the specific angles reveal basic air movement characteristics of each design.

TABLE 12.1 Air Movement Symbols

h	High velocity	n	No air movement
m	Medium velocity		Primary airflow
l	Low velocity		Secondary air currents
vl	Very low velocity	⊕	Compass direction
el	Extremely low velocity		Main airflow direction

The characteristics are about the same for both the true angle and the slightly skewed angles of air movement. In some test studies, the 45° angle had basically the same airflow patterns within the building as the 90° angle. When that was not true, the 45° angle is illustrated. Sections of each design are taken through the main living spaces. The resultant airflow patterns were analyzed and evaluated as to their optimum performance potential. Based on that information, the residential designs were slightly modified by the author, for experimental purposes only, to improve air movement. The modified designs were tested in the smoke chamber, and the resultant airflow patterns were analyzed and evaluated. (Table 12.2).

TABLE 12.2 Room Designation Symbols

E	Entry	S	Study
LR	Living room	AT	Atrium
FR	Family room	M	Music room
K	Kitchen	GR	Garden room
DR	Dining room	F	Fireplace
N	Nook	U	Utility room
P	Pantry	ST	Storage
MBR	Master bedroom	C	Closet
BR	Bedroom	T	Terrace or patio
MB	Master bathroom	G	Garage
B	Bathroom		

Residence One

A house located in New Canaan, Connecticut, was originally designed by Marcel Breuer in 1951. In 1976 a new couple acquired the house and 5 years later the couple enlarged the house while maintaining its original aesthetics. The long, horizontal, ground-hugging house is accented with massive stone walls, stone-laid terraces, and floor-to-ceiling windows as evidence of its modernist aesthetics. In short, the house is a landmark of architectural history, although it has been updated to meet the needs of the current owners.

The house was selected for analysis and evaluation for two reasons. First, the house was designed and built in the early 1950s. The performance of the house with regard to air movement control in comparison to houses of the late 1970s and early 1980s may reveal some interesting facts about residential designs. Second, the house has been modified on the interior, and an addition has been attached to it. The results of those improvements may yield some positive or negative effects in air movement control (Figures 12.1 to 12.13).

The house performs weakly as far as air movement control is concerned. Optimum airflow throughout the house is difficult to obtain, especially in the bedroom areas. The bathrooms would probably require mechanically forced air movement at least 60 percent of the time. Furthermore, the addition interferes with the potential opportunity of the main house to gain more air movement through the building. More openings in exterior walls permit airflow without much lost velocity and airflow patterns of greater benefit to the occupants. The relocation of a couple of doors also assists in allowing the airflow to travel with less hindrance inside the house. In short, several minor modifications, properly located, could significantly improve the capability of the house to control air movement.

Residence Two

William G. Wagner designed a typical project house for developers as an example of utilizing passive energy techniques in Florida.[1] The design was based on knowledge of local climatic conditions and basic principles of orientation, shading, heating, and ventilation. Mr. Wagner analyzed the site's climate with regard to wind, humidity, and temperatures. From those data, he made several observations. He developed specific energy-saving design priorities which evolved into a residential design form. His intent was to create a house plan that saved energy passively without being unusual or expensive to construct.

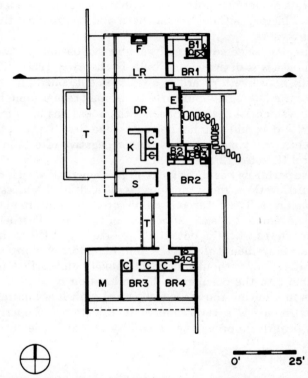

Figure 12.1 The interior of the New Canaan house was virtually gutted in 1981 and reconstructed with larger and improved spaces. Also, a bedroom wing was added to the existing house.

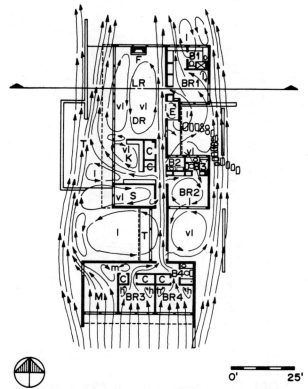

Figure 12.2 A northward airflow creates minimal air movement in the kitchen, study, bedroom 1, bedroom 2, bathroom 1, and bathroom 3. Most of the living room and dining room receives little air movement, and moderate airflow occurs along the peripheries. There is no movement of air in bathroom 2.

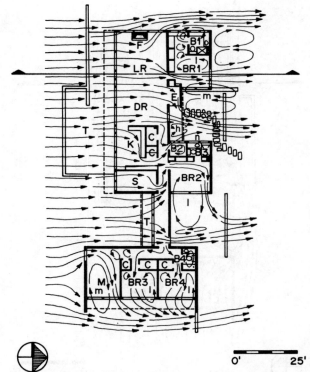

Figure 12.3 An eastward movement of air establishes little airflow in any of the bedrooms except bedroom 4. In bathroom 1 and bathroom 4, little air movement occurs; in bathroom 3, no air movement occurs. On the other hand, the main living spaces receive adequate airflow. Note the high-speed air swirl created in bathroom 2.

0'   25'

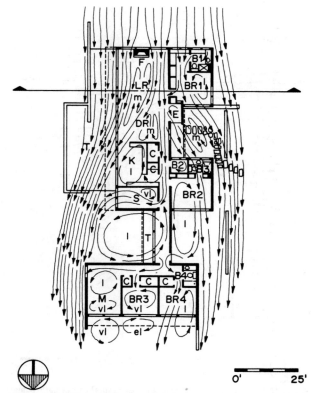

Figure 12.4 When the air movement is southward, most of the spaces receive good air movement except bedroom 3 and bathroom 4. Part of the study has adequate air movement; a small portion of the space does not.

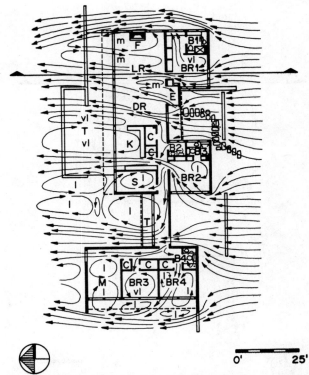

Figure 12.5 When the air movement is westward, the three bathrooms in the main house have no air movement while the kitchen, music room, bedroom 3, and bathroom 4 have only minimal air movement. Only the living room, dining room, study, bedroom 2, and bedroom 4 appear to receive a sufficient flow of air.

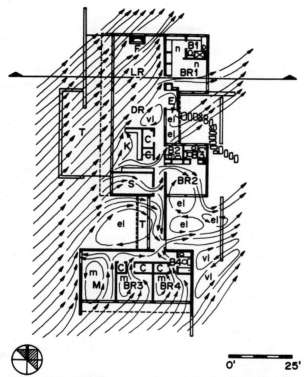

Figure 12.6 A northeast movement of air provides inadequate airflow in several interior spaces. The four bathrooms, bedroom 1, and bedroom 2 gain no significant benefits from the general air movement.

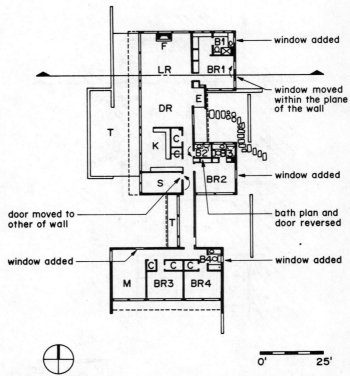

Figure 12.7 Several modifications to the New Canaan house aid in redirecting the airflow patterns into paths which appear to be more beneficial to the occupants.

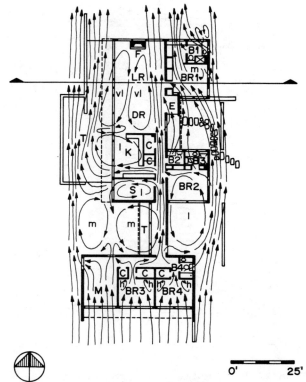

Figure 12.8 Openings added to the music room and bathroom 4 allow the air movement through those spaces to exit at an increased velocity. That, along with an additional opening in bedroom 2 and relocated doors in the study and bathroom 2, accelerates the airflow velocity in the general area. The relocated window in bedroom 1 and an opening added to bathroom 1 increase air movement in both those spaces.

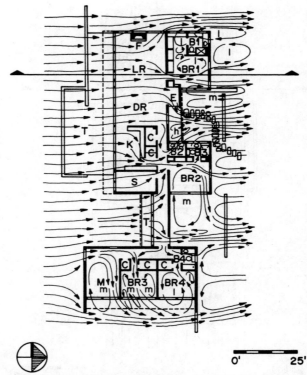

Figure 12.9 Relocating the doors in the study and bathroom 2 and adding an opening in bedroom 2 increase the movement of air in all three spaces as well as in bathroom 3. Although the airflow patterns in bathroom 1 and bedroom 1 are altered, the overall velocity of the airflow is relatively unchanged. Adding an opening in bathroom 4 increases the flow of air within that space and changes both the velocity and direction of the airflow in bedroom 4 while increasing only the air movement velocity in bedroom 3.

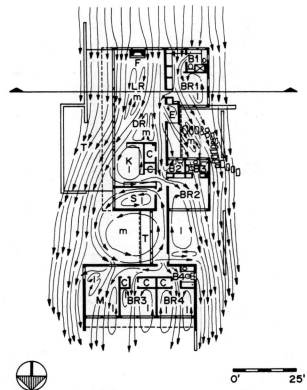

Figure 12.10 Adding an opening in the music room and bathroom 4 increases the flow of air in both rooms and also in bedroom 3. Relocating the study door accelerates air movement within that space. The remaining spaces of the house already receive adequate movement of air regardless of the other building modifications.

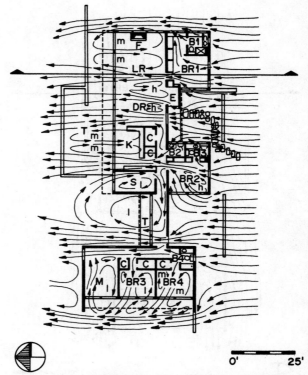

Figure 12.11 Properly located openings in the house increase the flow of air throughout a greater portion of the house, especially in all four bathrooms and bedrooms as well as the kitchen, study, and music room.

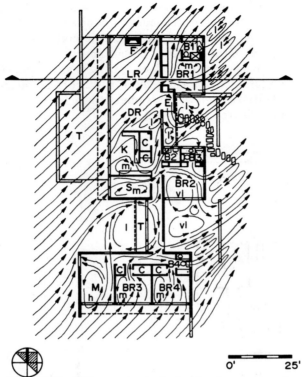

Figure 12.12 Adding openings and relocating doors improve the movement of air throughout a greater portion of the house, especially in all four bathrooms, bedroom 1, bedroom 2, music room, and dining room.

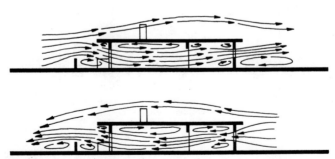

Figure 12.13 Sections of the New Canaan house exhibit excellent airflow patterns. The general movement of air seems to occur mainly in the living zones of the house regardless of the airflow direction. When the air travels in an eastward direction, the terrace wall appears to aid in guiding the airstream downward into the occupant area.

Selection of the house for analysis and evaluation was due to its departure from conventional houses. The house was oriented to the south rather than to the street, which reduced the heat load on the building by having the smaller portions of the house facing east and west. In addition, several outdoor activity spaces were established rather than creating extra indoor spaces such as a family room. However, Mr. Wagner did suggest that this particular residential design was only one of the many possible designs for the specific site (Figures 12.14 to 12.29).

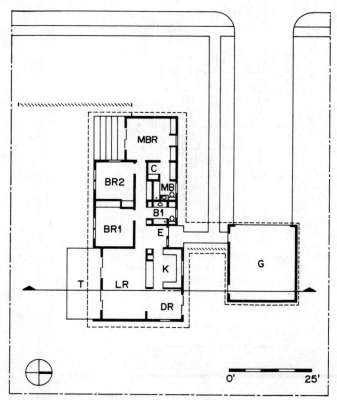

Figure 12.14   A floor plan based on such climatic concerns as orientation, shading, air movement, and solar gain. (*Adapted from Ref. 1*)

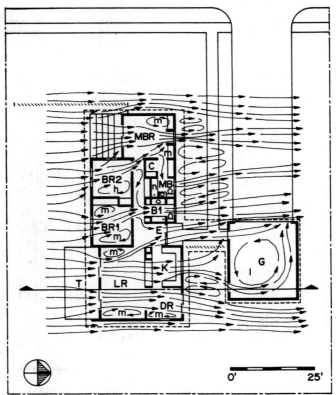

0'   25'

Figure 12.15 Northward movement of air provides excellent airflow in most rooms except the master bedroom, where no air movement occurs.

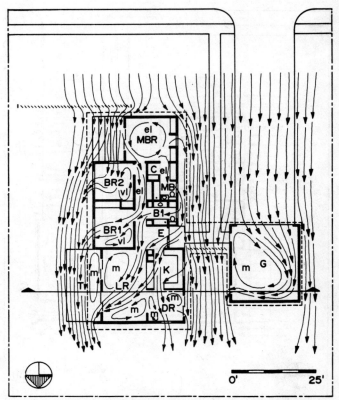

Figure 12.16 An eastward airflow creates minimal air movement in all the bedrooms and the master bathroom. Only the living room, dining room, kitchen, and entry receive adequate air movement.

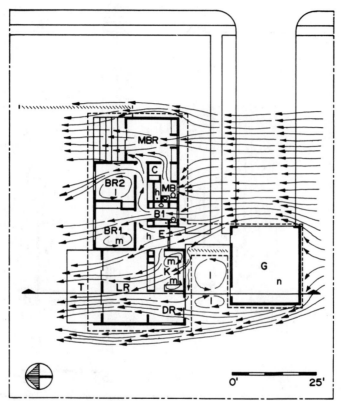

Figure 12.17 When the air movement travels to the south, all of the rooms receive adequate airflow. Note the space between the garage and the kitchen. Air movement is brought into the kitchen by the eddy, which is an unusual occurrence.

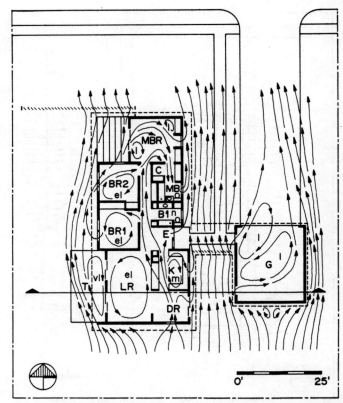

Figure 12.18 Most of the house receives little or no air movement when airflow is westward. Only the dining room, kitchen, and entry have adequate air movement.

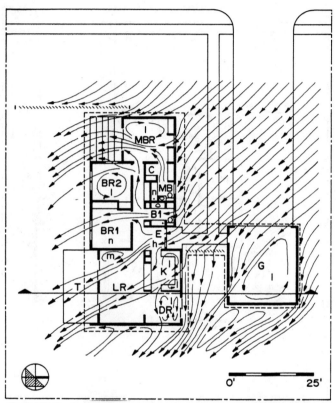

Figure 12.19 Southeasterly air movement provides adequate airflow in all the spaces except bedroom 1, where no air movement occurs. The movement of air in the master bedroom is usually low because of room location. The reason is easily spotted: the opening in the north wall of the bedroom is too small for the angled airflow to enter the space, and the garage usually redirects the main air movement away from it.

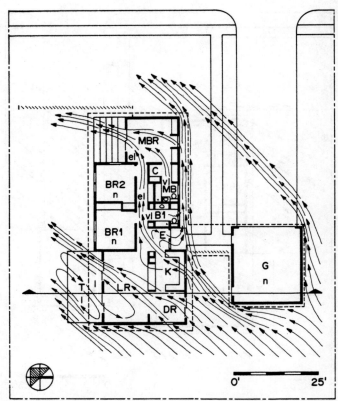

Figure 12.20 When the air movement travels southwesterly, most of the house receives minimal airflow. All of the bedrooms and bathrooms obtain inadequate air movement while the kitchen, dining room, living room, and entry receive good air movement. The garage appears to block most of the airflow from reaching the house.

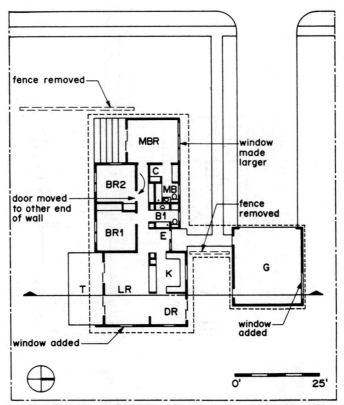

fence removed

window
made
larger

MBR

C

BR2

MB

door moved
to other end
of wall

B1

BR1

E

fence
removed

T

LR

K

G

DR

window
added

window added

0'   25'

Figure 12.21 Minor modifications of the house aid in redirecting the airflow patterns into more beneficial paths with respect to the occupants of the dwelling.

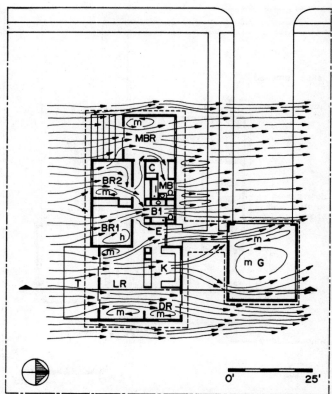

Figure 12.22 The enlargement of the north opening in the master bedroom allows more air to travel through the space with a slight increase in velocity and helps to provide adequate air movement in the master bathroom. In the remainder of the house, no extensive improvements are required to supply sufficient air movement.

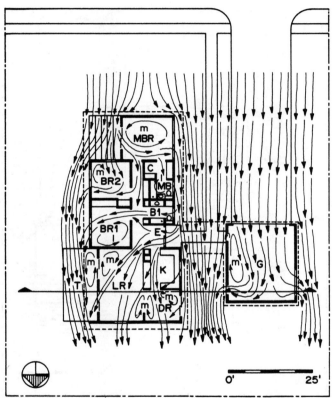

Figure 12.23 The addition of an opening in the living room and the enlargement of the opening of the master bedroom provide needed exits for the flow of air through the house. Air movement in all three bedrooms and the master bedroom is sigificantly improved.

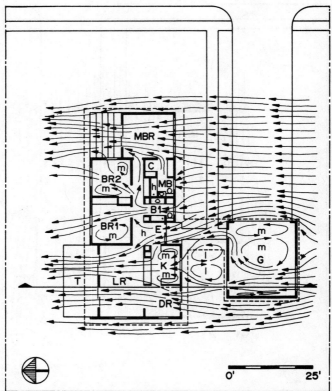

Figure 12.24 The introduction of an opening in the north wall of the garage allows air movement to occur within the house. Enlarging the opening in the master bedroom permits increased airflow within that space. The relocated door in bedroom 2 and the velocity increase in the master bedroom result in improved air movement in bedroom 2.

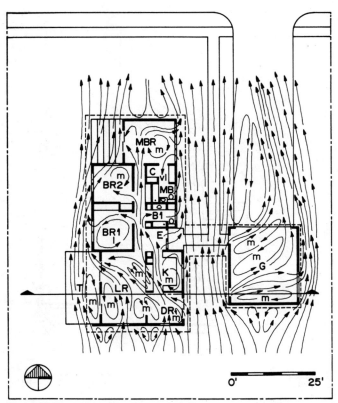

Figure 12.25 Adding an opening to the living room and enlarging the north opening in the master bedroom increase the movement of air throughout the entire house.

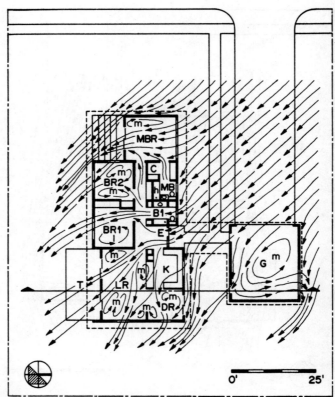

Figure 12.26 Air movement in the dining room is improved by adding an opening in the north wall of the garage. Enlarging the opening in the master bedroom increases the airflow velocity within that space. Relocating the door in bedroom 2 results in improved air movement in bedroom 2 and bedroom 1 by enlarging the inlet-to-outlet ratio.

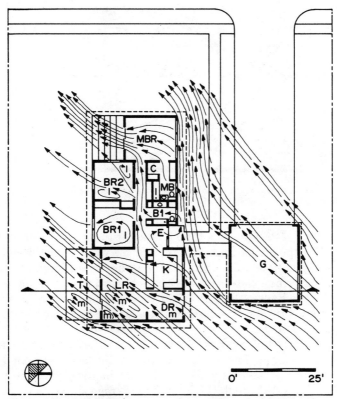

Figure 12.27 Adding an opening in the north side of the garage improves the overall airflow within the house by allowing the movement of air to reach the building in the first place. Enlarging the opening in the master bedroom and relocating the door in bedroom 2 assist in increasing the air movement velocity in all three bedrooms and both bathrooms. The added opening to the living room allows more airflow within that space.

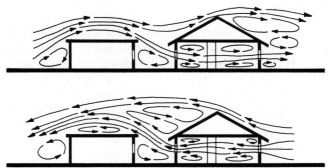

Figure 12.28 Sections of the house and garage reveal the interference that the garage poses with respect to the airflow.

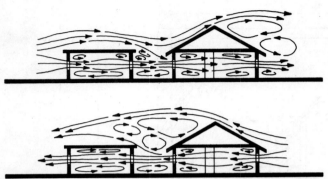

Figure 12.29 Openings in the garage provide thorough air movement within both the house and the garage. The airflow occurs at the occupant level in each interior space regardless of air movement direction.

The residential design has a dual character in regard to air movement control. When the air movement is toward the west or east, the long axis of the house, the flow of air through the building is extremely poor. On the other hand, when the air movement is toward the north or south, the short axis of the house, the flow of air through the building is excellent. Even when air movement is toward the southeast or southwest, airflow within the house is fair and very poor, respectively.

The critical areas of weak air movement are the bedrooms and bathrooms. The interior airflow is corrected by relocating a bedroom door and adding openings to several walls, especially in the garage and master bedroom. A section through the house and garage shows that the garage interferes with airflow through the house. The modifications made to Mr. Wagner's residential design assist in providing better air movement control.

Residence Three

The Harris house, in Yukon, Oklahoma, was designed by Richard A. Kuhlman.[3] This early 1950 house was developed to be self-cooling based on two simple ideas. First, a properly designed house may be kept reasonably cool during the summer months in a hot climate without air conditioning. Second, the house could be cooled more economically and effi ciently with air conditioning if the designer used climatically oriented design ideas. "The Harris Home has been called the only house in town with a breeze because it is so designed and located that it actually creates a breeze where apparently none had existed."[2]

The Harris house was selected for analysis and evaluation because of its architect's concern for energy consumption. Mr. Kuhlman utilized a ventilated roof, air scoops, venturi effect, reflective roof, and central court. The house was designed to work with the climate instead of against it (Figure 12.30 to 12.41).

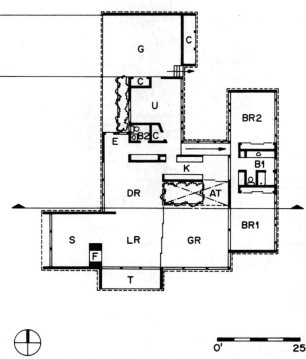

Figure 12.30 The Harris house incorporates a central open court which allows air to move through. (*Adapted from Ref. 2*)

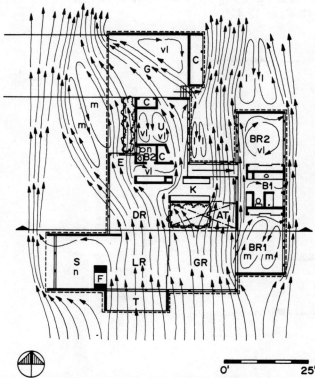

Figure 12.31 A northward movement of air provides adequate airflow in a majority of the rooms. In the utility room and bedroom 2, air movement is not sufficient to create comfort. No airflow occurs in bathroom 2 or in most of the study.

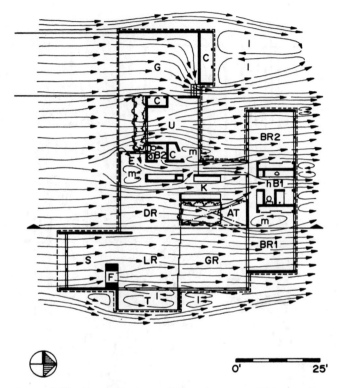

Figure 12.32 Air movement traveling to the east establishes excellent airflow in all rooms of the house.

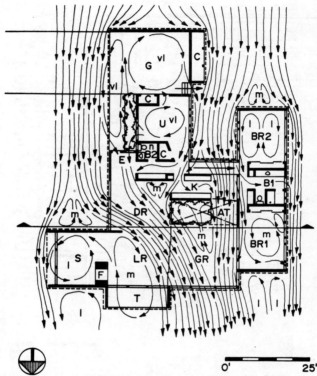

Figure 12.33 A southward movement of air creates good airflow through most of the house. The utility room and garage receive inadequate air movement, and bathroom 2 gains no movement of air.

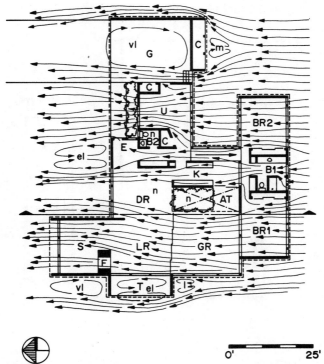

Figure 12.34 A westward airflow creates air movement through most of the house except in the dining room and central court.

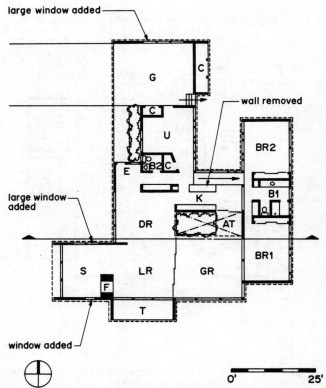

large window added

wall removed

large window added

window added

G

C

C

U

E

B2 C

BR2

B1

K

DR

AT

BR1

S

LR

GR

F

T

0'   25'

Figure 12.35 A few modifications made to the Harris house improved the movement of air throughout the house, which benefits the occupants.

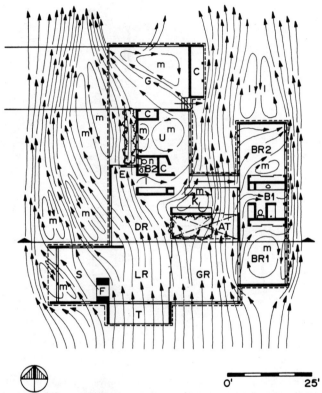

Figure 12.36 Adding openings in both the north and south walls of the study greatly increases the movement of air through the study. The velocity of the airflow in the utility room, garage, kitchen, and bedroom 2 has been improved by removing the kitchen wall.

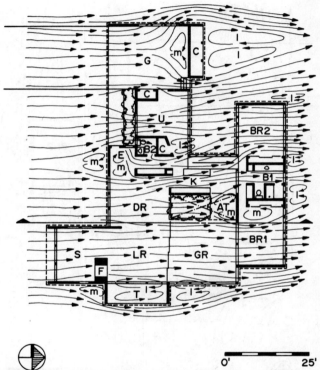

Figure 12.37 Although no major improvement can be made in this particular situation with regard to air movement through the house, adding an opening in the garage has slightly increased the flow of air within that space.

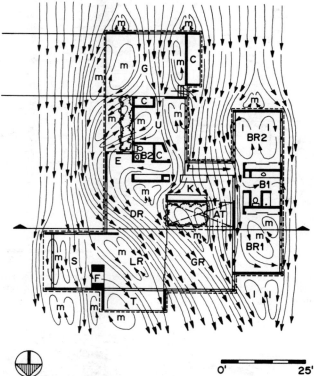

Figure 12.38 Adding an opening in the north wall of the garage has improved the movement of air within the utility room, entry, dining room, kitchen, and garage. The airflow through the kitchen is further enhanced by removing the kitchen wall. Openings added in both walls of the study have siginifcantly increased the flow of air within the space.

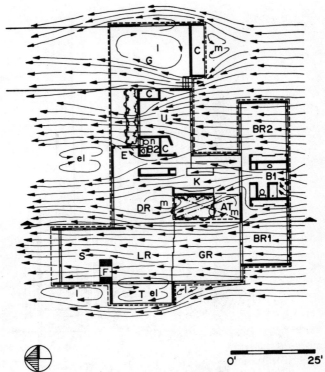

Figure 12.39 Removing the kitchen wall allows the flow of air to increase velocity within the area and create air movement in the central court and dining room.

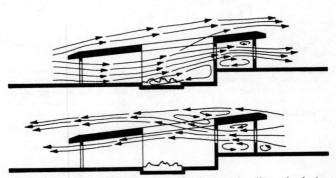

Figure 12.40 Sections of the Harris house illustrate the effects of a sloping roof and changes in the ground plane. Eastward air movement travels through the house at the occupant level while westward air movement exits up through the roof or moves along the ceiling.

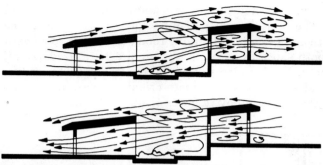

Figure 12.41   Height reductions of the window and door openings cause the airflow, eastward or westward, to travel through the house at the level of the occupants.

Air movement throughout the Harris house is quite good. The utility room and bathroom 2 are the only two spaces whose airflow is not similar to that in the remainder of the house. The flow of air within the utility room is slightly lower in velocity, and the bathroom receives no air movement 80 percent of the time. Bedroom 2, on the other hand, obtains good airflow all the time except in one specific situation when the airflow is northbound.

The Harris house is modified by the addition of a few openings and the removal of a kitchen wall. The slight changes increase the velocity of the air movement through several spaces. Only bathroom 2 does not benefit from the minor changes; some serious alterations would be needed to provide that space with air movement. A section through the house illustrates the airflow pattern within the interior spaces. When the airflow is eastward, the streams travel along the occupant level; but when the airflow is westward, the streams exit through the central court or move along the ceiling plane. The reduction of opening heights draws the airflow into the occupant level regardless of its direction. The movement of air through the Harris house could be only slightly improved because of its well-thought-out plan.

Residence Four

The architectural firm of Rowes Holmes Associates designed the Logan house for the climate of Tampa, Florida.[3] The house was patterned after the traditional southern dog trot homes of early Florida. The plan is very open in the center portion of the house. The clients claim they have used the mechanical heating and cooling equipment rarely and yet manage to stay very comfortable.

The Logan house was selected for analysis and evaluation because of its simple design. The house consists of nine squares. Located in the middle three squares are the kitchen, dining room, and living room with private rooms on both sides. In addition, the architects utilized several energy-saving ideas, including clerestory windows, a belvedere, ceiling fans, tin roof, and raised floors. The climate-oriented house promotes a casual style which the clients enjoy (Figures 12.42 to 12.55).

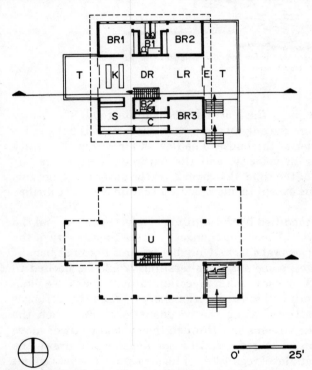

Figure 12.42 The Logan house revitalizes the traditional dog trot houses of the South. (*Adapted from Ref. 3*)

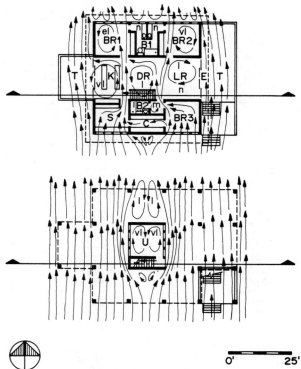

Figure 12.43 A northward flow of air at the living level establishes good air movement in the study and bedroom 3, fair air movement in the dining room, living room, and bathroom 2, inadequate air movement in the kitchen, bedroom 1, bedroom 2, and no air movement in bathroom 1. The same northward airflow at the ground level provides poor air movement in the utility room while the surrounding areas receive excellent air movement.

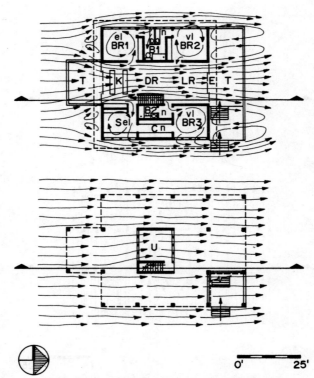

Figure 12.44 An eastward movement of air provides good airflow in the kitchen, dining room, and living room. On the other hand, the remaining spaces receive inadequate or no air movement. Throughout the entire ground level, the eastward air movement creates an excellent flow of air.

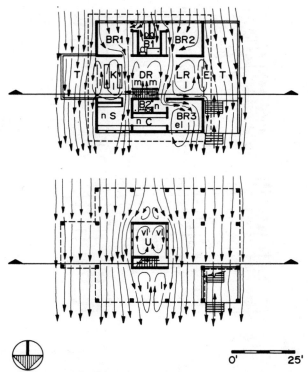

0'   25'

Figure 12.45 At the living level, southward air movement creates good airflow in bedroom 1, bedroom 2, bathroom 1, and the dining room, and adequate airflow in the kitchen and living room. The study, bathroom 2, and closet obtain no air movement while bedroom 3 receives minimal air movement. The southward airflow at ground level establishes inadequate air movement in the utility room while excellent air movement occurs in the surrounding area.

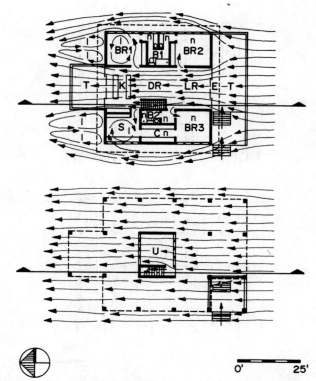

Figure 12.46 At the living level, westward air movement creates no airflow in bedroom 2, bedroom 3, bathroom 1, bathroom 2, and the closet while a low level of airflow is achieved in bedroom 1 and the study. However, the kitchen, dining room, and living room are excellent examples of good air movement. The westward movement of air provides adequate airflow throughout the entire ground plane.

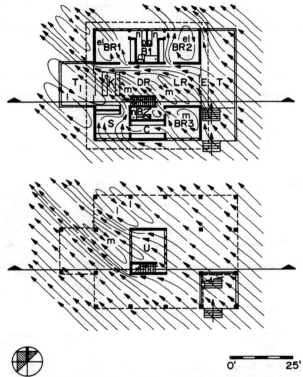

Figure 12.47 Northwest air movement creates good airflow in the kitchen, dining room, living room, study, and bedroom 3. Neither bathroom receives air movement, and bedrooms 1 and 2 obtain inadequate air movement.

O'   25'

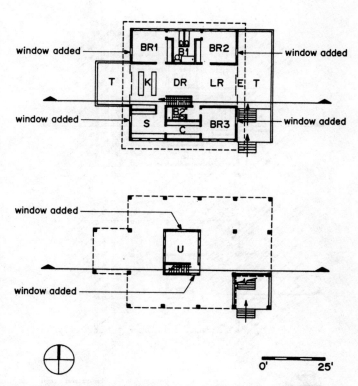

window added

window added

window added

window added

window added

window added

0'   25'

Figure 12.48 Several modifications to the Logan house at both the living and ground levels aid in improving the movement of air within the house.

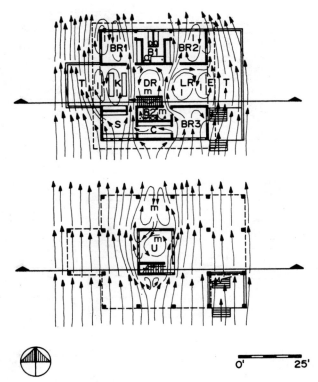

0'   25'

Figure 12.49 Openings added to the living level increase the movement of air within the kitchen, dining room, bedroom 1, and bedroom 2, but no improvement is made in bathroom 1. Openings added at the ground level significantly improve the movement of air in the utility room.

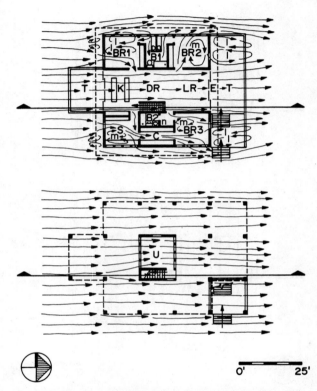

Figure 12.50 Openings added to the living level in the bedrooms and study improve the movement of air throughout the entire level except in bathroom 2. At the ground level, air movement cannot be significantly improved in this particular situation.

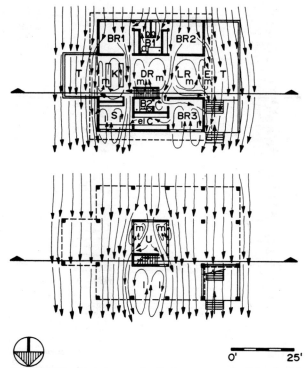

Figure 12.51 Air movement in all the spaces at the living level is greatly improved by adding openings in the three bedrooms and study. At the ground level, openings increase the movement of air in the utility room.

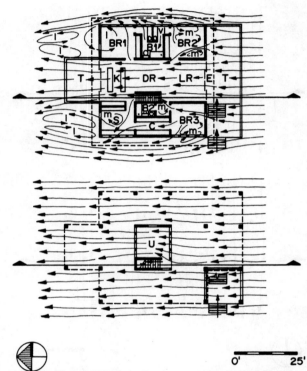

Figure 12.52 At the living level, openings added to each bedroom and the study increase the overall velocity of air movement throughout the house, especially in those particular rooms. Air movement at the ground level in this specific situation cannot be significantly improved.

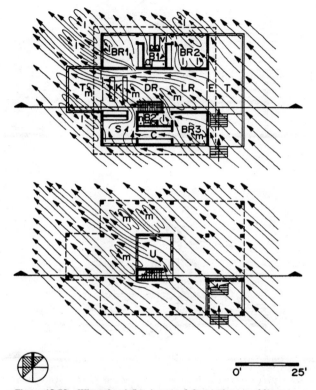

Figure 12.53 When the airflow is toward the northwest, adding openings at the living level increases the movement of air within that level except in bathroom 2.

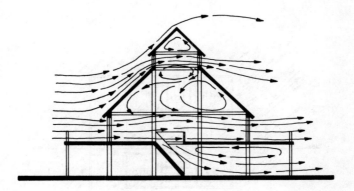

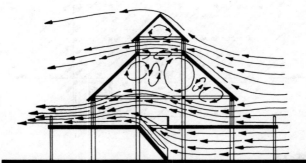

Figure 12.54 Sections of the Logan house show that the roof form has little effect on the movement of air. In both directions, air movement occurs in the occupant zone at the living level. Notice the slight jump in the airflow as additional air movement is introduced into the living level.

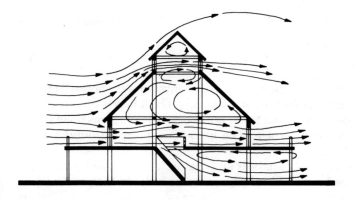

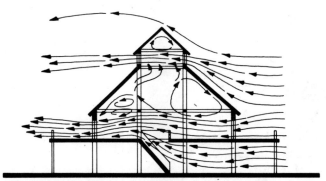

Figure 12.55 The reduction of the door heights concentrates the airflow in the occupant zone at the living level.

The Logan house responds very poorly to proper air movement control. At best, only 60 percent of the house receives adequate airflow, whereas 30 percent is not uncommon. The private rooms suffer most from inadequate air movement. The simple addition of four openings at the living level improves the movement of air through the house significantly. With those additions, up to 90 percent of the building receives good airflow 100 percent of the time. A section through the middle of the house shows that the primary airflow occurs at the occupant level. However, westbound air movement creates an upward jump in its pattern because of the flow of air coming up the stairwell. The reduction of opening heights concentrates the airflow more at the occupant level and decreases the jump effect.

Residence Five

Home Planners, Inc., developed a one-story contemporary house featured in their annual magazine, *Houses and Plans.*[4] The house is zoned for efficiency, privacy, and easy access to the outdoors. All the rooms open to the rear onto an ideal party terrace. The house is centered around a large living room.

Selection of the house for analysis and evaluation was based on its typical contractor-type layout. The design is based only on the proximity of spaces. It does not respond to site conditions or climatic factors (Figures 12.56 to 12.68).

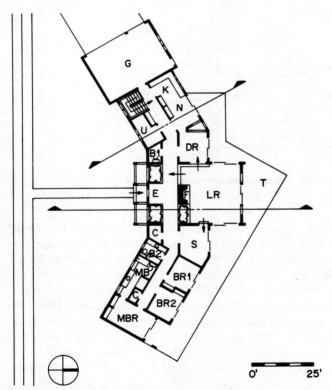

Figure 12.56 This one-story contemporary house centers around a large living room. (*Adapted from Ref. 4*)

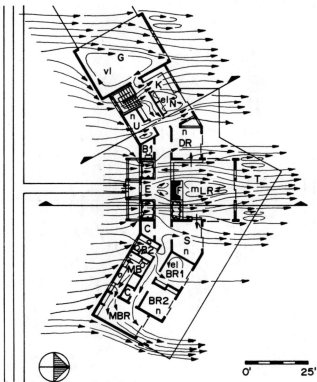

Figure 12.57 A northward movement of air creates minimal airflow in a large number of the rooms. Bedroom 2 receives no air movement. Parts of the study, utility room, and dining room gain no air movement. Portions of the kitchen, nook, garage, and bedroom receive inadequate air movement.

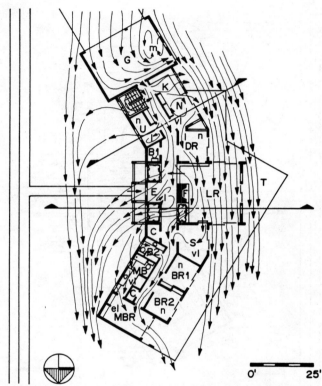

Figure 12.58 Eastward air movement establishes poor airflow throughout the house. Parts of the dining room and utility room receive no air movement, nor do bedrooms 1 and 2. The kitchen, nook, bedroom 3, master bedroom, and bathroom 2 obtain inadequate airflow.

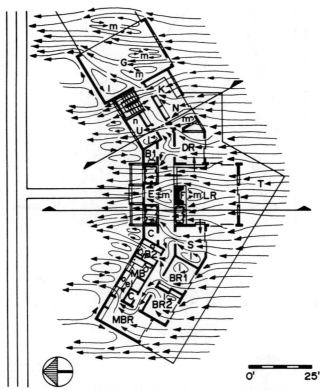

Figure 12.59 A southward flow of air provides good air movement in most of the rooms except the master bathroom and part of the utility room.

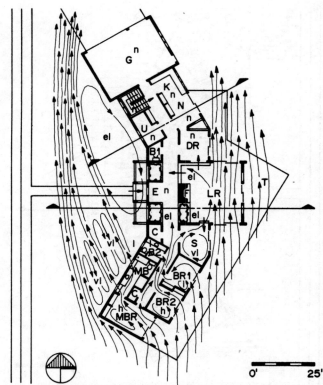

Figure 12.60 Most of the house receives little or no air movement when airflow is westward. Only the bedrooms, bathroom 2, and master bathroom obtain adequate movement of air.

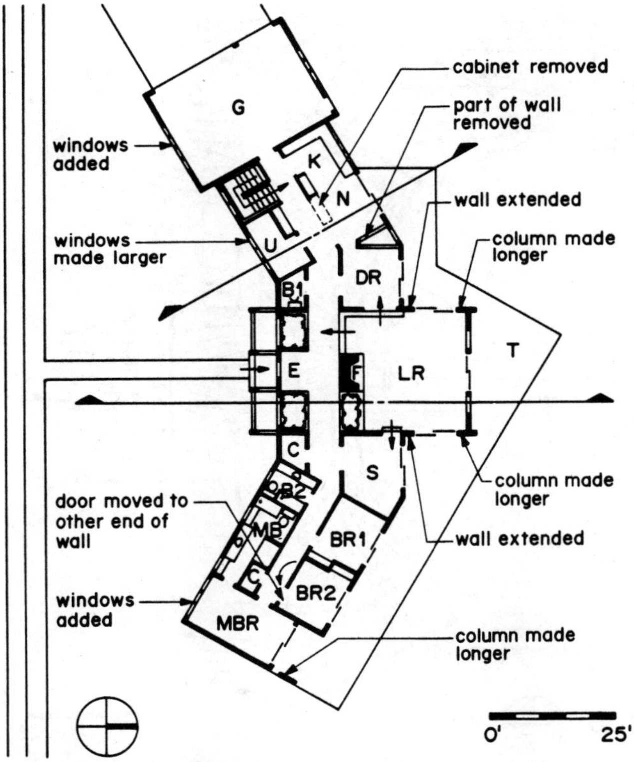

G
K
N
U
DR
B1
E
F
LR
T
C
S
B2
MB
BR1
C
BR2
MBR
windows added
windows made larger
cabinet removed
part of wall removed
wall extended
column made longer
column made longer
wall extended
door moved to other end of wall
windows added
column made longer
0'
25'

Figure 12.61 Several modifications of the house aid in redirecting the airflow patterns into more beneficial paths with respect to the occupants.

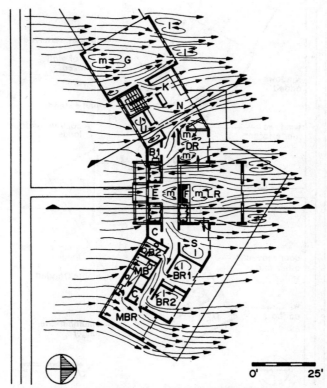

Figure 12.62 Adding an opening in the garage improves air movement in both the garage and kitchen. Enlarging the opening in the utility room and removing extra kitchen cabinet space increase the flow of air in the nook, utlity room, and part of the dining room. Adding an opening in the master bedroom and relocating the door in bedroom 2 improve the movement of air in all three bedrooms.

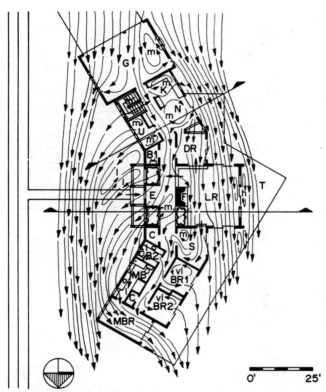

Figure 12.63 Adding an opening in the garage improves the movement of air throughout the house. Although significant improvements are made in bedrooms 1 and 2 with regard to air movement due to the additional opening in the master bedroom, the opening location in this particular situation limits the degree of potential improvement.

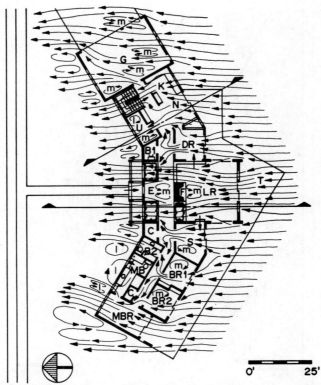

Figure 12.64 Enlarging the opening in the utility room allows more air movement through the space. Adding a large opening in the south wall of the master bedroom increases the airflow in the master bathroom, bedroom 1, bedroom 2, and the master bedroom. Relocating the door in bedroom 2 also aids in improving that room's air movement.

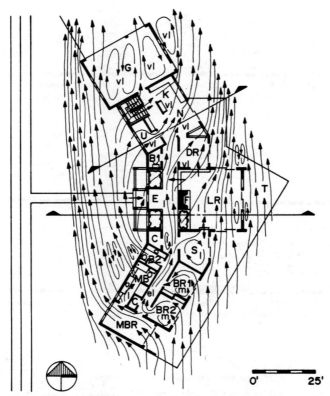

Figure 12.65 Adding a large opening in the master bedroom increases air movement within that particular space. That and adding an opening in the garage and extending walls and columns in the living room assist in improving air movement quality throughout the entire house.

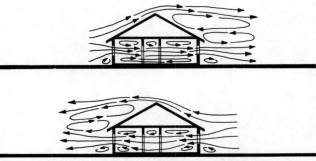

Figure 12.66 Sections of the house indicate that the major portion of the airflow occurs at the occupant level.

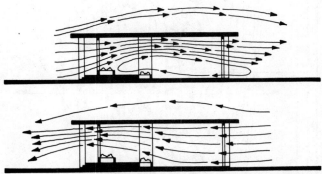

Figure 12.67 Sections of the house reveal the upward motion of airflow as it travels through the interior space regardless of its direction.

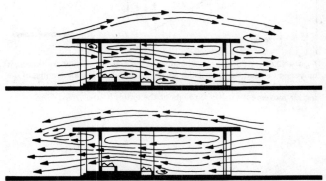

Figure 12.68 Reducing heights of the door and window openings causes the airflow pattern to travel through the occupant level.

The residential design responds to air movement in three different ways. Flow of air through the building is extremely weak when the exterior air movement is westward or eastward. Only 30 percent of the house receives any significant air movement, and air movement in the remaining 70 percent is seriously inadequate. Northward air movement fares only slightly better; 60 percent of the house obtains minimal air movement. However, good airflow is obtained through most of the interior spaces when the air movement is southward. Several modifications made to the plan of the house, such as new openings, enlarged openings, relocated openings, and extended structural elements, improve the flow of air within the building.

The efficiency of the northward and southward air movement increases significantly, whereas westward and eastward airflows improve only moderately. Air movement in the latter two directions may be greatly improved by making tremendous alterations in the building form. Sections of the house are taken through the bedroom and living room areas. In the bedroom area, the air flows at the occupant level, and in the living room area, the airflow direction is upward toward the ceiling. The reduction of opening heights in the living room area brings the airflow into the occupant zone.

Major Weaknesses

The design of residential buildings has not significantly improved with regard to air movement control in decades; a house built in the 1950s is just as effective as a house designed in the 1980s. Four major points of weakness are evident in most residential designs. First, the houses have a directional characteristic. The form of the building creates poorer air movement in one direction than in another. Residence five is an excellent example.

Second, detached garages or wings interfere with or block the flow of oncoming air. The separate building may stop the movement of air from reaching the main house by deflecting the airflow. Or the separate building may redirect or guide the flow of air into specific patterns which are not beneficial to the main house. The solution is to attach the building to the main house and add openings to it, skew the detached building in relation to the main house, or change garages into carports and building wings into screened porches.

Third, bathrooms seem to be problem areas so far as air movement is concerned. It is important to provide openings at two different points for minimum air movement.

Fourth, because the openings are usually too great in height, they allow an upward movement of airflow within thebuilding. The reduction of opening heights assists in concentrating air movement at the occupant level.

From analysis and evaluation of the five residential designs, an innate incompatibility is seen to exist between them and air movement. Regardless of how climate- and energy-conscious an architectural designer was, the building form does not seem to maximize or optimize the potential ability of the house to control air movement. In short, most residential designs are lacking in strong capability to guide and redirect airflow patterns to benefit the occupants.

REFERENCES

1. The Bureau of Research, *Houses and Climate: An Energy Perspective for Florida Builders,* The Governor's Energy Office, Tallahassee, Fla., 1979.
2. "20% Cooler without Air Conditioning," *House and Home,* February 1955.
3. David Morton, " 'Dog Trot' House," *Progressive Architecture,* vol. 62, June 1981.
4. "One-Story Contemporary Houses," Home Planners, Inc., Farmington, Mich., 1982.

13

Techniques Applied
to the Design Process

Architectural designers can produce buildings which passively remain cool in hot weather and warm in cold weather, without the use of mechanical systems, when the structures are designed from their inception to enhance climatic benefits and reduce climatic extremes. Control of air movement is one variable that exerts a great influence on climatic comfort within a building. Consequently, a structure based on economy and function cannot reasonably be expected to have a quality indoor climate unless attention is given to climatic issues and air movement control. No architectural designer can develop a climatically successful building unless the design is fundamentally responsive to climatic concerns. Air movement control must be accounted for when the basic concept of the design is developed.

Design is a multiple-problem-solving situation. The architectural designer must simultaneously resolve the issues of energy, climate, topography, vegetation, mechanical systems, structural systems, economy, operation, function, sociality, psychology, and physiology. He should not lose sight of any of those issues while coordinating them into a functioning system within a specific context. The application of air movement control techniques involves all the design issues. However, it is impossible to establish a universal order of priorities in dealing with those issues because the degree of importance varies from one design to another. One of the purposes of the design development stage is to analyze the issues, determine the priorities, and develop a coherent and integrated design solution. The finished design is always evaluated by the complete solution produced.

If buildings are to be designed climatically correctly, the architectural designer should have a knowledge and understanding of the techniques of air movement control. Five separate residential designs are developed to illustrate the techniques. None of them are established as ideal for optimum air movement control, since an ideal house for every design situation is not possible and the site characteristics are not examined. The anticipated and actual airflow patterns of each design will be evaluated and discussed in detail.

Each residential design was submitted to smoke chamber tests, and the resultant airflow patterns were analyzed and compared to anticipated patterns of air movement. The angles of attack, the angles of direction of airflow with respect to the plane of the building side facing the airflow, utilized in these studies were 90° and 45° (Table 13.1). Although actual air movement rarely approaches a building at a true 90° or 45° angle, those angles reveal basic air movement characteristics of each design which are relatively accurate for both the true and slightly skewed angles of air movement. Sections of each building are taken through the main living spaces. The resultant airflow patterns are analyzed and evaluated as to their optimum performance potential (Table 13.2).

TABLE 13.1 Air Movement Symbols

h	High velocity	n	No air movement
m	Medium velocity	→	Primary air flow
l	Low velocity	⧂	Secondary air currents
vl	Very low velocity	⊕	Compass direction
el	Extremely low velocity	▰	Main airflow direction

TABLE 13.2 Room Designation Symbols

E	Entry	S	Study
LR	Livingroom	AT	Atrium
FR	Family room	M	Music room
K	Kitchen	GR	Garden room
DR	Dining room	F	Fireplace
N	Nook	U	Utility room
P	Pantry	ST	Storage
MBR	Master bedroom	C	Closet
BR	Bedroom	T	Terrace/patio
MB	Master bathroom	G	Garage
B	Bathroom		

Design One

The house was developed as a simple rectangular box, which is typical of most conventional homes. However, several unconventional features were incorporated into the design. The doors leading to the bedrooms open into one central area of the hall rather than at different points along the hallway. A bathroom also opens to the same central area. This feature aids the movement of air through the bedroom area with less hindrance than in a conventional house. In addition, the living room, dining room, and kitchen flow openly into each other, which permits air movement to travel easily through all three spaces. The house is a blend of conventional and unconventional residential design ideas (Figures 13.1 to 13.10).

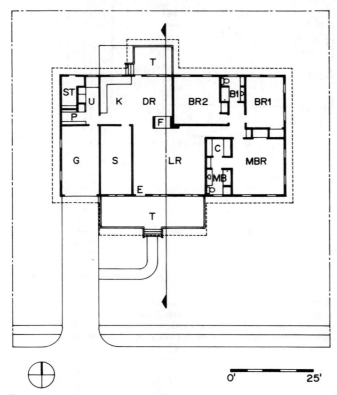

Figure 13.1   This house design utilizes a conventional rectangular form as a point of departure for some unconventional ideas.

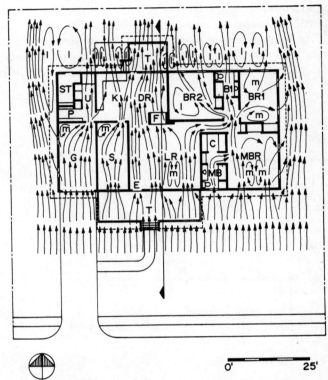

Figure 13.2 Northward air movement establishes good airflow throughout the house. In the corner of bedroom 2, air movement is slightly low but is quite able to produce comfort. A similar low area of air movement occurs in bedroom 1. The reverse flow of air in bedroom 2 was unexpected; however, it does provide adequate air movement.

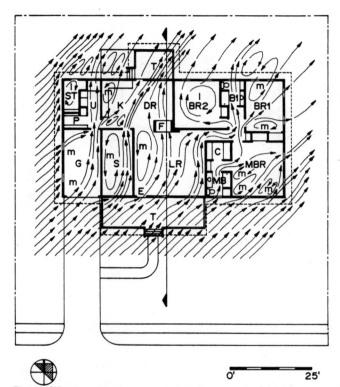

Figure 13.3 A northeast movement of air creates excellent air movement within all the spaces of the house. Only the airflow in bedroom 2 is less than in the remainder of the house. Note how the negative pressure zone beside the utility room establishes air movement inside the storage room, an unpredicted occurrence.

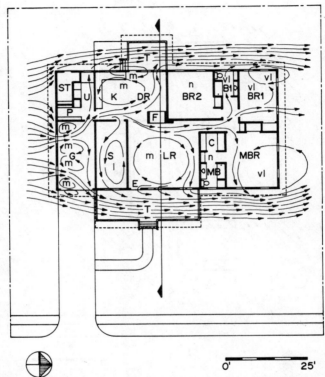

Figure 13.4 Eastward air movement provides good airflow within the living room, dining room, and kitchen. However, the bedrooms and bathrooms receive minimal or no air movement. The fallacy is in the longitudinal layout of the house, a typical characteristic of conventional homes, which causes one space to block airflow from reaching another space.

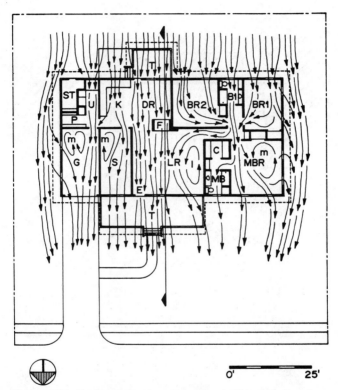

Figure 13.5 Southeasterly air movement establishes good airflow throughout the entire house. Secondary air movement occurs in areas where the direct airflow does not occur.

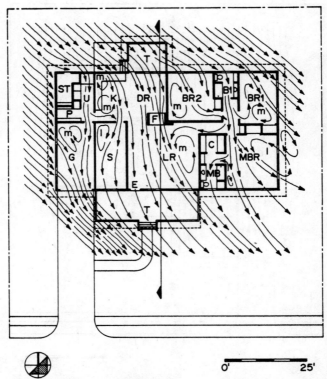

Figure 13.6 Southbound air movement provides excellent airflow throughout the house. In areas where direct airflow does not occur, secondary air movement is established.

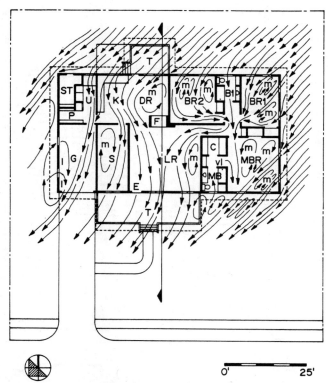

Figure 13.7 Southwesterly air movement creates good airflow within all the rooms except the master bathroom. The back flow of air movement in the garage, bedroom 1, and master bathroom was not anticipated. The airflow within those spaces was expected to travel in the same direction as the main exterior airflow.

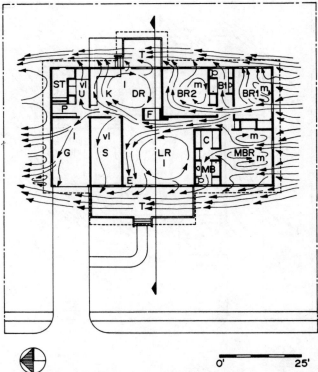

Figure 13.8 Westward air movement establishes a good flow of air in the house. However, the study and utility room received inadequate air movement. The longitudinal layout of the house and the location of the door in the study hinder the air movement from satisfactorily entering the study.

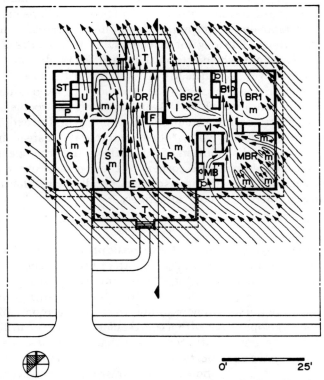

Figure 13.9 A northwest movement of air provides good airflow within the interior spaces of the house.

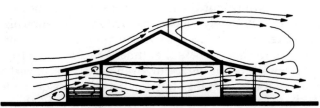

Figure 13.10 Whether the air movement is southward or northward in direction, a section of the house reveals that the airflow dips downward while traveling through the occupant zone. This positive effect is negated when the outside railings of the porches are removed. The airflow then moves through the interior space above the occupant level.

The house performs well with respect to air movement control. Although the eastbound and westbound airflow directions create some areas of poor air movement, the remainder of the house receives more than an adequate amount. The longitudinal layout of the residence is the main source of this problem, which is typical in most, if not all, conventionally built houses. However, the difficulty is not as severe because of the openness of the main living spaces and the adjacency of the bedroom doors. In addition, the ease with which the airflow enters the rooms is a direct result of the amount and placement of the openings in the exterior walls. Notice that the openings are not grouped together but instead are spread apart for a thorough distribution of the airflow. Also, the openings, like the doors, are located near the corners of the rooms in order to assure complete air movement within the spaces. Overall, the residential design utilizes air movement relatively effectively.

Design Two

The house was designed according to two basic principles. First, the layout of the residence is linear. Second, the depth of the residence is kept to a minimum of two rooms except for a bathroom. The bedrooms open to a central hall in relatively the same area, which allows air to move easily from room to room. The kitchen, dining room, living room, and family room are one large open area with a central closet. Note the utilization of two openings in the master bathroom rather than the usual single opening. The garage also plays an important role in the effectiveness of gaining air movement control in the house. The house employs several design features which may assist the movement of air within its interior spaces (Figures 13.11 to 13.20).

Air movement throughout the house is excellent except for one minor flaw. The garage completely blocks the airflow from entering the main house at one time, and it reduces the velocity of the airflow from reaching a small portion of the house at another time. The solution is to add several openings in the walls of the garage or, even better, to convert the garage into a carport and add openings in the adjacent walls of the house. The cluster effect of the bedroom doors and the openness of the main living spaces appear to assist the movement or air within the house. The introduction of two openings in the master bathroom allows secondary movement of air when direct airflow is not available to the space. The residential design's utilization of linear layout and minimum depth as design limitations provides good air movement.

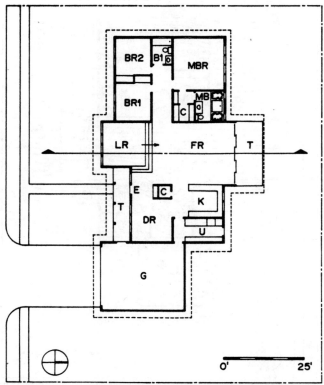

Figure 13.11 The house utilizes linear length and minimal depth as design elements in order to obtain good air movement control.

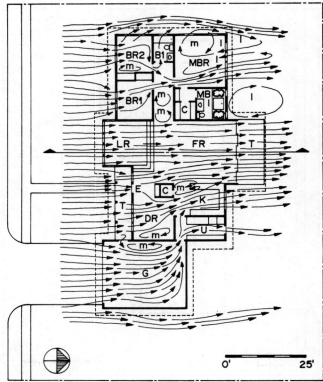

Figure 13.12 Northward air movement creates excellent airflow through all the rooms. Notice the circular airflow patterns in the master bedroom, master bathroom, and hall.

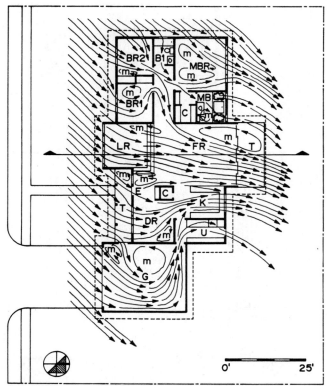

Figure 13.13 Northeasterly air movement establishes superb airflow throughout the entire house. No areas of poor air movement are found.

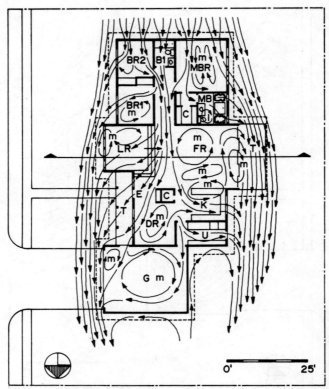

Figure 13.14 Eastward air movement creates excellent airflow within all the spaces of the house including the garage.

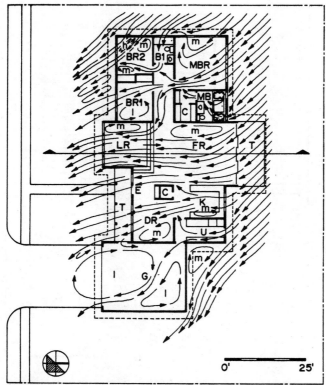

Figure 13.15 A southeasterly movement of air provides excellent airflow throughout the house. The reverse flow of air in the master bedroom was not anticipated; it is caused by the velocity of the airflow entering the north opening.

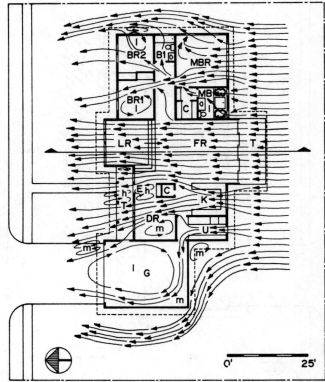

Figure 13.16 Southward air movement establishes excellent airflow throughout the house.

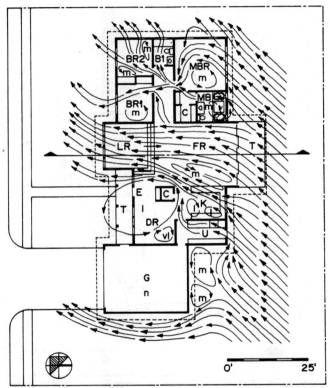

Figure 13.17 Southwesterly air movement creates good airflow within all the spaces of the house except the dining room, where it is slightly less. The reduced flow of air in the dining room is due to the location of the garage. If the garage were a carport instead and openings were placed in the west wall of the dining room, excellent airflow would occur within that space. The master bathroom receives air movement that is due to the swirling nature of the airflow.

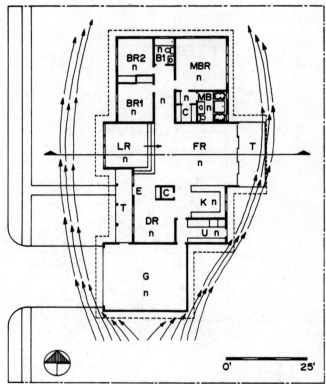

Figure 13.18 With westward airflow, the garage does not allow the house to receive any air movement. The airflow through the house would be significantly increased if the garage were converted to a carport.

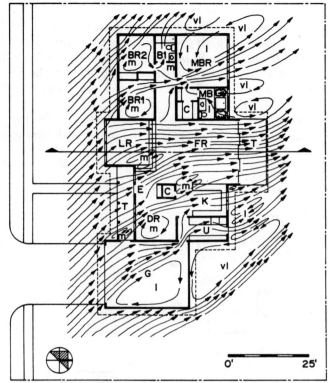

Figure 13.19 A northwest movement of air provides excellent airflow within the house. The master bedroom receives a good flow of air in part of the space while the remaining area obtains a lower movement of air. The master bathroom gains air movement which is lower in velocity than in the remainder of the house.

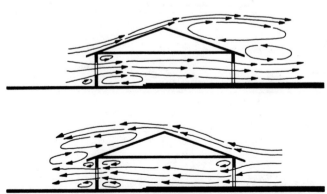

Figure 13.20 Regardless of air movement direction, the flow of air as it passes through the house is directed into the occupant level.

Design Three

The house was established by the clustering of spaces. The layout of the rooms creates areas of both protrusions and indentations. The garage is semidetached from the main house as a means of reducing its impact on air movement within. The play of solids and voids in the house design may prove to be beneficial with regard to air movement control (Figures 13.21 to 13.30).

The house controls air movement quite successfully. The only problem with acquiring complete airflow within the house is the garage. It needs

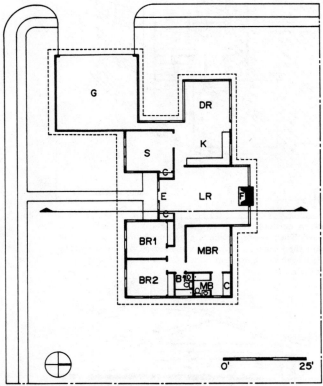

Figure 13.21 The clustered plan of the house allows variety in the shape, size, and orientation of individual spaces while maintaining the potential of successful air movement control.

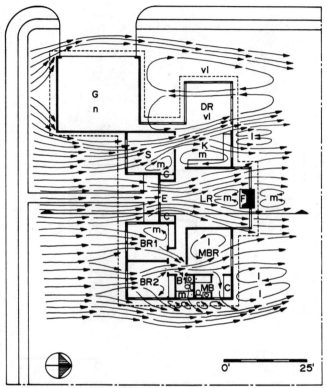

Figure 13.22 Northward air movement creates good airflow within all the rooms except the dining room. The flow of air in the dining room is inadequate to provide comfort. Note the unusual swirl patterns of the secondary air movement currents in the master bathroom.

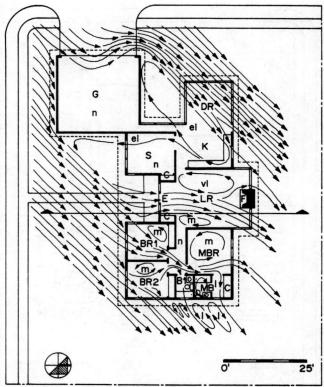

Figure 13.23 A northeasterly movement of air establishes good airflow in over half of the house while the remainder receives inadequate airflow. The situation may be quickly remedied by adding openings in the garage or by converting the garage into a carport. The master bathroom receives secondary air movement by swirling currents.

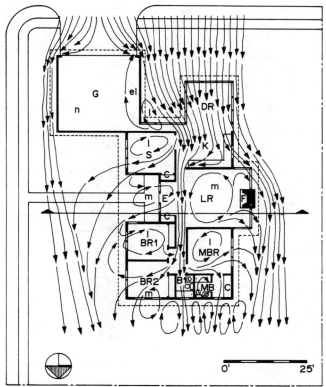

Figure 13.24 Eastward air movement provides excellent airflow throughout the house. While the garage increases the flow of air in the dining room and kitchen, it reduces the potential velocity of the air movement in the study, bedroom 1, and bedroom 2. The flow of air in the master bathroom is inadequate to provide comfort.

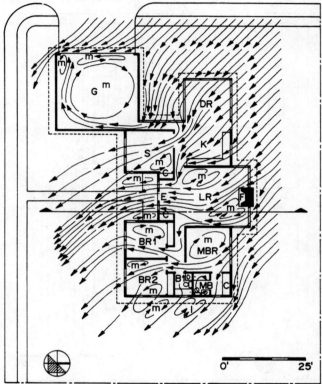

Figure 13.25 Southeasterly air movement creates superb airflow within all spaces of the house. The garage aids in increasing the velocity of the airflow through the study.

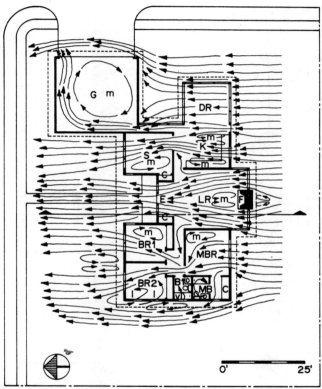

Figure 13.26 Southward air movement establishes good airflow in all the main living spaces. However, the unanticipated direction of the airflow in the two bathrooms is inadequate for the provision of comfort.

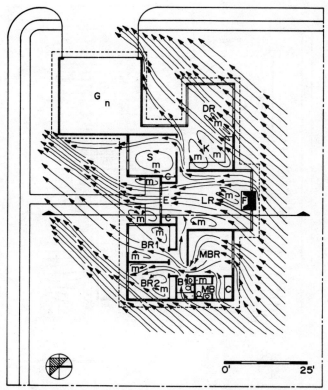

Figure 13.27 A southwesterly movement of air provides excellent airflow throughout the entire house. Note the reverse airflow pattern in the master bathroom, an unpredicted occurrence.

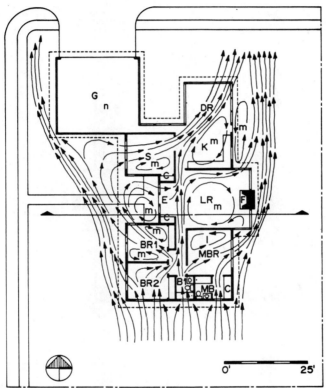

Figure 13.28 Westward air movement creates good airflow within the interior spaces. Several secondary currents of air movement occur in locations where direct airflow does not.

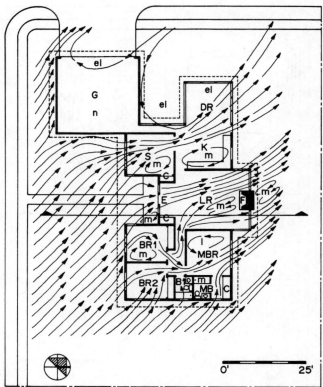

Figure 13.29 Northwesterly air movement provides thorough airflow inside the house. An area of inadequate air movement occurs only in part of the dining room, and no air movement is detected in the garage.

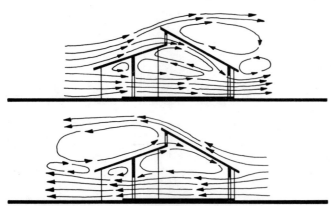

Figure 13.30 The airflow travels through the occupant level regardless of the direction of the exterior air movement. This is achieved by limiting the height of the openings in the outside walls. The upper portion of the interior space receives air movement, which may reduce the impact of solar heat gain.

to be altered. Either adding openings to the garage or transforming the garage into a carport is necessary. The layout of the master bathroom may be reversed to facilitate improved airflow within the space, although the bedroom does receive adequate air movement most of the time. Secondary air movement currents are helpful in establishing good airflow throughout the house.

Design Four

The house design utilizes a central atrium surrounded by living spaces. The plan is nondirectional, an extreme departure from most conventional homes. The plan is also extremely open, which allows the interior spaces to flow from one to another much like the anticipated airflow. The atrium may be screened over, or operable insulated windows may be installed for more climatic control of the space. The garage is strategically located to assist in good air movement control and to reduce its potential for hindering the flow of air. The unconventional characteristics may improve the capability of the house to control air movement successfully (Figures 13.31 to 13.41).

The house establishes excellent control over the movement of air. No major flaws, or even minor ones, could be detected in the study model. The house performed as well as expected on a large scale, and better than expected on a small scale.

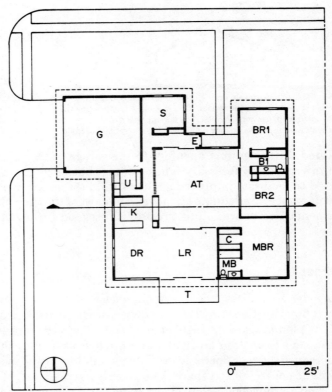

Figure 13.31 The atrium plan of the house permits interaction between the spaces and allows individual rooms to focus both inwardly and outwardly.

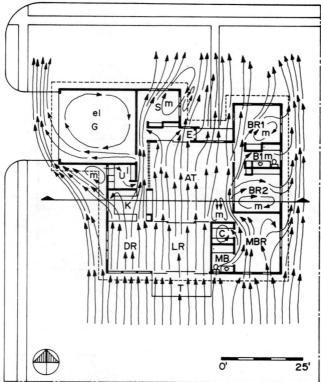

Figure 13.32 Northward air movement provides excellent airflow in all the rooms in the house. Over 90 percent of the interior space receives direct air movement.

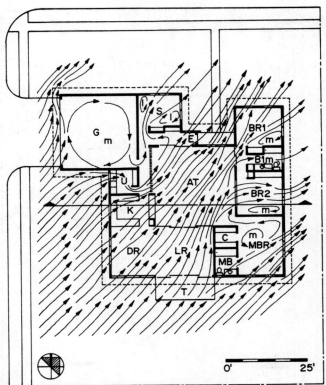

Figure 13.33 A northeasterly movement of air establishes superb airflow within all the spaces of the house. Secondary air currents are kept to a minimum.

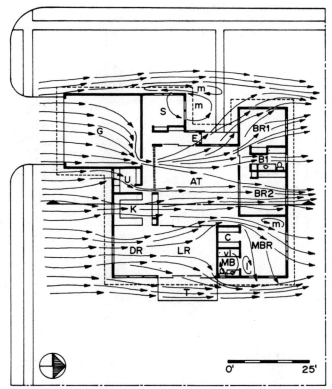

Figure 13.34 Eastward air movement establishes good airflow within all the spaces of the house. Only the master bathroom receives airflow inadequate for the provision of comfort.

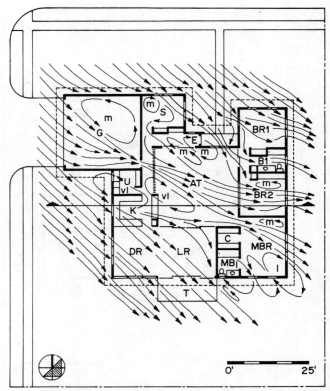

Figure 13.35 Southeasterly air movement provides a good flow of air throughout the house. The utility room and a small portion of the atrium obtain poor airflow. Secondary air currents assist the direct flow of air in successfully creating optimum air movement within the house.

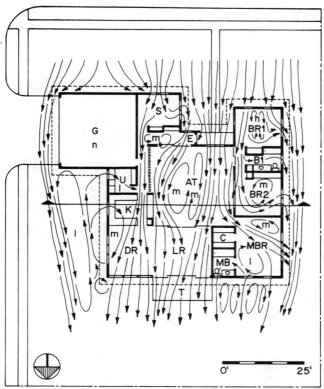

Figure 13.36 Southward air movement creates excellent airflow inside the house. Even the large secondary air current in the atrium establishes a comfortable movement of air.

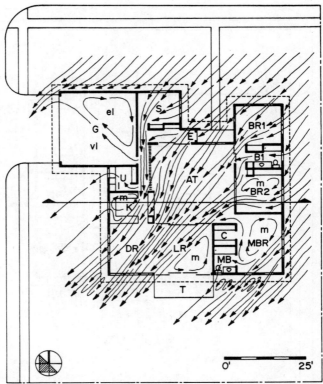

Figure 13.37 A southwesterly movement of air establishes thorough airflow within all the rooms. Notice the small number of secondary air currents in the house, which indicates that most of the building receives direct air movement.

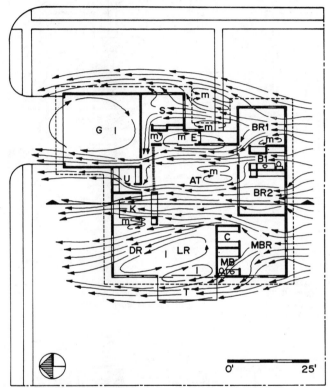

Figure 13.38 Most of the house obtains superb airflow with a westward movement of air. Although most of the living room receives air movement due to secondary currents, the flow of air is more than adequate to provide comfort.

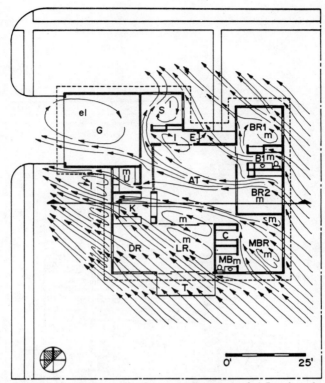

Figure 13.39 Northwesterly air movement creates good airflow throughout the entire house. Secondary air currents provide most of the airflow for the living room quite effectively.

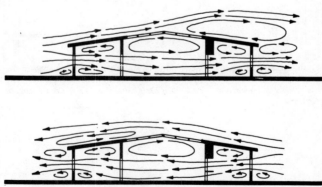

Figure 13.40 The flow of air through the house occurs at the occupant level regardless of the outside direction of air movement. Whether the atrium is screened, glazed, or roofed, the airflow travels right over the roof. An open atrium allows a small volume of airflow to spill into its space; however, the airflow is likely to bring excess heat into the atrium from the roof.

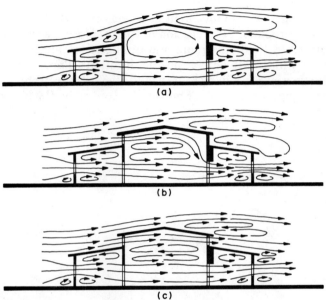

Figure 13.41   Three variations of building section provide some interesting results in air movement control for this type of house plan. With the atrium roof raised only (a), the secondary air currents near the ceiling only involve more area. When an opening is introduced just below the raised roof (b), air movement is permitted to enter the atrium and be brought down into the occupant level. When two openings are made just below the raised roof (c), the airflow travels through only the upper portion of the atrium, which removes heated air collected below the ceiling.

Design Five

The house was developed by utilizing an angular configuration. The main living spaces make up a central core, with the bedrooms and study establishing the wings. The wings were kept to a depth of one room while maintaining the capability of each space to open freely to the outside. The unusual plan may assist the house in obtaining beneficial air movement control (Figures 13.42 to 13.49).

The house controls the movement of air relatively well with one major exception, the fireplace. The location of the fireplace creates an unusual airflow pattern in the living room and dining room, and it reduces the airflow velocity in the kitchen in specific situations. The relocation of the fireplace further into the living room would alleviate the problem. However, the overall performance of air movement control by the house is quite satisfactory.

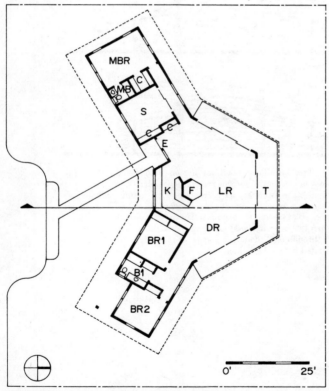

Figure 13.42   This house plan employs a scheme using a central core with wings to provide privacy for the bedroom areas.

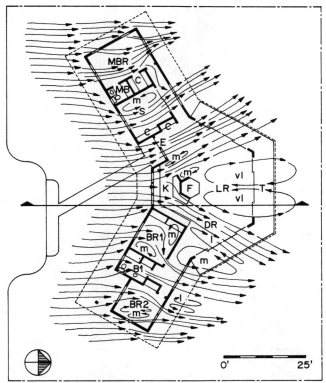

Figure 13.43 Northward air movement provides excellent airflow throughout the house with one unusual occurrence. Note the airflow pattern in the living room and dining room. The fireplace causes the airflow to shoot off to each side of those spaces. That occurred whether the fireplace was hexagonal or rectangular in shape. The reason for the unanticipated happening is the close placement of the fireplace to the opening in the kitchen with respect to living room size. Eliminating the fireplace or reducing openings in the living room would solve the situation presented. Northwesterly and northeasterly movements of air create this same airflow pattern.

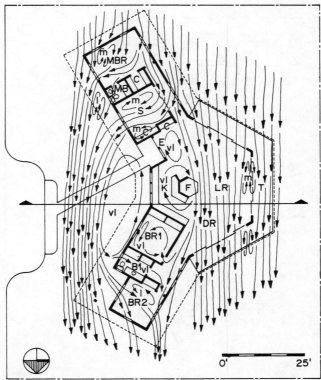

Figure 13.44 Air movement to the east establishes good airflow within the interior spaces of the house. However, the kitchen, bedroom 1, and bathroom 1 receive only minimal airflow. The condition may be improved by moving the fireplace more into the living room.

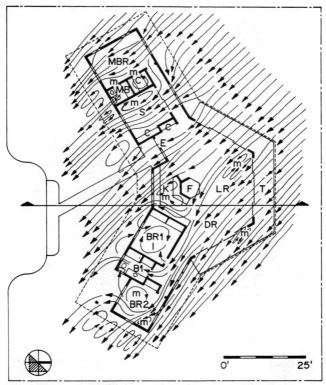

Figure 13.45 Southeasterly air movement provides good airflow in each room. The two east wing bedrooms obtain good air movement through secondary currents. Note the flow reversal in bathroom 1 created by bedroom 2.

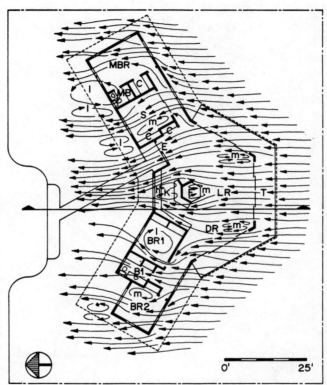

Figure 13.46 Southward air movement provides good airflow throughout the house. The kitchen receives accelerated airflow, and the master bathroom and bedroom 1 receive low air movement created by secondary currents.

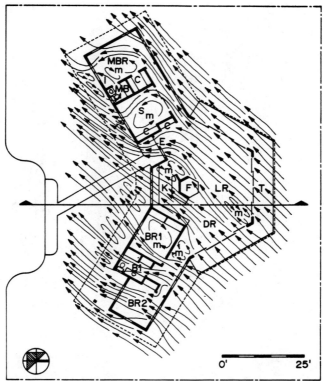

Figure 13.47 A southwesterly movement of air establishes excellent airflow within all the spaces of the house. Secondary air currents are created in the study, bedroom 1, master bedroom, and master bathroom.

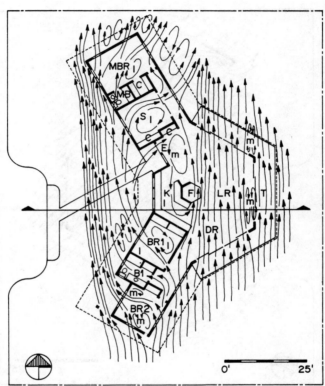

Figure 13.48 Air movement traveling westward creates good airflow in each room, although about half of the movement of air is caused by secondary currents.

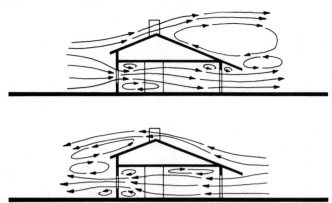

Figure 13.49 Regardless of its direction of movement, the airflow travels through the interior space at the occupant level. However, note the increased velocity of the airflow when it is southward. It is caused by the small inlet opening in the kitchen and the large outlet opening in the living room. The reverse occurs when the airflow is northward.

Design Considerations

Residential buildings which are climatically responsive are designed with the consideration of the effects of proper air movement control as in the five designs. Several specific features of the designs appear to be relevant to the provision of successful control of air movement.

First, interior doorways of adjacent spaces should open into a common area, such as a hallway, to optimize airflow through those spaces. Second, spaces which are abutting and do not need to be closed off may flow freely from one to another in order to maximize the movement of air within the spaces. The living room, family room, dining room, and kitchen are such spaces. In other words, one large space may be divided into different activity zones.

Third, bathrooms need at least two openings, although three openings are better, to achieve adequate air movement. Fourth, each space should have a minimum of two exterior openings which are not adjacent. Fifth, the placement of garages needs careful consideration. An incorrectly located garage may seriously hamper the movement of air within the house. If a garage is not really necessary, the designer may consider a carport. Residential designs, as illustrated, have the potential capability of controlling air movement effectively.

14

Final Remarks

Architectural designers who create buildings that are climatically correct have developed a basic understanding of air movement control within particular microclimatic environments. Although the task is complex, techniques may be utilized to assist in the control of air movement. The techniques enlarged upon in this guide are not cookbook recipes to be followed step by step; they are tools for decision making—the foundation for understanding the methods of air movement control. They aid architectural designers in making intelligent and reasonable climatic decisions within the design process. The anticipated result of the application of these techniques is that buildings will respond to their microclimatic environment by drawing positive benefits from their surroundings.

Air movement control offers a method of improving the quality of life within residential structures. Three benefits are obtained. First, the air quality in residences is significantly improved through the dispersion and removal of smoke, gases, toxins, odors, and other contaminants. Second, residential energy consumption and operating costs are lowered by the reduction of heat gain and the acceleration of heat loss during hot months, and vice versa during cold months. Third, superior human comfort, both physical and mental, is provided through increased convection and body evaporation rates. The lack of air movement may have the reverse effect in all three situations. Therefore, permitting and accentuating the controlled movement of air allows nature to assist designers in achieving acceptable human comfort levels in their buildings.

Architectural designers need to have a knowledge and an understanding of the important aspects of architecture if building forms are to be effective and accurate in controlling air movement. First, designers should be familiar with the basic characteristics of air movement. Second, an understanding of the techniques of air movement control would be helpful. Third, information about the site and its surroundings, such as topography, landscaping, microclimate, fences, adjacent structures, bodies of water, and air movement patterns, should be gathered.

Fourth, designers need to analyze the airflow paths within the schematic design of the building. The analysis should reveal the general airflow patterns through and around the building, the location of eddies, and the distribution of the air pressure differentials. Fifth, the schematic design should be inserted into the site's climatic environment. The building's form needs to be reanalyzed and modified to conform to its surroundings, or the surroundings may be slightly altered to assure proper air movement control. Sixth, the final building design may be established by designers on the basis of the results of the analysis. Through these six steps, architectural designers may obtain the benefits of proper air movement control.

The exploration of air movement control techniques has progressed in a very unique manner. During the 1950s and early 1960s, the study of air movement was at its peak, although air conditioning had made the home scene in the 1930s. By the middle 1960s, over 25 percent of American homes had air conditioning, and the percentage was rapidly growing because of low power rates. Consequently, the concern for air movement control dwindled until the late 1970s, when the costs of electricity soared quickly and the public became concerned that fossil fuel supplies were becoming limited.

At that time, an interest in air movement control was reborn in both the United States and England. Mathematics and wind tunnels were the methods of air movement studies in the 1950s and early 1960s, and during the 1970s the computer joined the exploration team. However, since the 1950s, no new discoveries or insights have made a great impact on the theories of air movement control. At the present time, the wind tunnel still seems to be the best way to study air movement with respect to economy, flexibility, accuracy, variety, and details. The advancement of air movement control exploration has been a matter of slow progression.

To date, no single presentation of information on air movement control, written or oral, has covered all the possible site characteristics, building configurations, opening variations, and other design modifications, nor has any presentation documented its various probabilities. Consequently, there is need for further exploration and research in the field. Architectural designers in general practice cannot spend large amounts of energy and time resolving the climatic needs of each and every design problem, nor can they start with assumptions about important features of each building with regard to air movement control. Instead, they need facts upon which to base intelligent decisions. This book unites the explorations of past researchers and develops a relatively comprehensive guide of known air movement control techniques.

Appendix

Smoke Chamber Testing

A kerosene smoke airflow chamber was used to investigate the movement of air through residential designs. The testing procedure, model construction, and operation of the smoke chamber are discussed. The documentation of the building forms used in the tests and the resultant airflow patterns functions as supportive evidence of the design issues involved in air movement control. The information and insight gained from the employment of the smoke chamber are discussed and incorporated in the main body of the book.

Procedure

Ten individual residential designs were tested in the kerosene smoke airflow chamber. Five of the designs were developed by architects. They were examined as to their potential of positive air movement control, which is the ability of the designs to guide and redirect airflow patterns to benefit the occupants. The design plans and sections were converted into test models. The models were inserted into the smoke chamber and tested. The resultant airflow patterns were analyzed and evaluated with regard to the original and modified designs. Five other designs were developed by the author to illustrate the techniques of air movement control in climatically designed residences. The plans and sections of the building forms were converted into test models. Again the models were tested. The anticipated and actual airflow patterns of each design were compared, analyzed, and evaluated. The results and conclusions of the tests are discussed in Chapters 12 and 13.

Figure A.1 A kerosene smoke airflow chamber was utilized to test Plexiglas models.

Test Model Construction

Each residential design was transferred into two test models. One illustrated the floor plan taken through the building at a height of 3 to 4 ft so that window and door openings were accounted for. The other illustrated a vertical section through the building showing the location and placement of window and door openings, ceiling heights, roof forms, and floor level changes. In one design, two sections were useful in studying building configurations. The test models were investigated in two dimensions, whereas the interpretation of the data was in three dimensions.

The test models are easily constructed when the following 10 steps are utilized. First, draw the plan of the house to scale. A scale of ⅟₁₆ in equals 1 ft is the minimum size that is usable in the smoke chamber. Second, cut the black mat board 1 in larger than the plan in both directions. Third, draw the plan onto the mat board with a pencil or a white colored pencil. Fourth, score the clear Plexiglas, several times, at 1-in intervals and break into 1-in strips. Pliers and gloves may be helpful. Leave the protective paper on the Plexiglas sheet and strips while working with them. Fifth, measure the lengths of the walls on the plan, and transfer the measurements onto the 1-in Plexiglas strips. Sixth, score the Plexiglas strips in the correct wall length sections, and break into the proper sizes. Remove the protective paper from the new pieces. Seventh, glue the pieces to the mat board in the correct locations by using plastic model cement. When two or more pieces abut, glue them together as well. Make certain that the glue completely fills the area between the two pieces; otherwise, the smoke stream may pass through the space between the pieces and cause inaccurate test results. Eighth, let the model dry for at least 1 hr. Ninth, drill two holes, ½ in apart, into the mat board. Use a ⅛-in drill bit and locate the holes near the center of the model. The

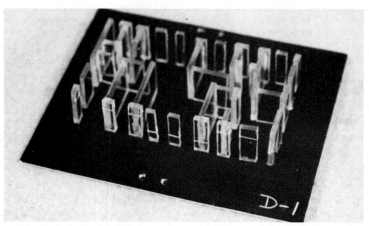

Figure A.2   Each residential design was converted into test models in both plans and sections.

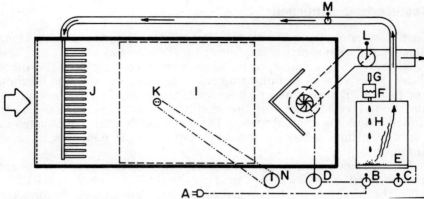

Figure A.3   Operation of the kerosene smoke airflow chamber involves eleven simple steps.

1. Plug the smoke chamber's AC plug (A) into a 120-volt AC outlet.
2. Turn on the main power switch (B) and the heater switch (C).
3. Turn the fan velocity down by turning the fan speed control knob (D).
4. Let the heater (E) warm for approximately 7 minutes. Meanwhile, follow steps 5 through 7.
5. Fill the kerosene regulator bottle (F) with kerosene. Make sure the kerosene regulator screw (G) is closed enough not to allow the kerosene to drop into the kerosene smoker bottle (H). Close the screw by turning it clockwise.
6. Open the back panel (I) and clean the smoke nozzles (J) with pipe cleaners so they are clear of deposits.
7. Place a test model on the model mount (K), and close the back panel (I). The orientation of the model may be changed by turning the angle-of-attack control knob (N).
8. Open the kerosene regulator screw (G) slightly by turning it counterclockwise until a frequency of about one drop per 1 to 2 seconds is obtained.
9. Open the exhaust valve (L). When the lever is moved completely to the right, as viewed from the glass side of the smoke chamber, the valve is open.
10. Open the smoke valve (M). When the lever is at the 12:00 o'clock position, as viewed from the glass side of the smoke chamber, the valve is open.
11. Start the fan circulating at high speed, 90 to 100 on the fan speed control knob (D), for 1 or 2 seconds. Then turn the velocity down, 65 to 75 on the fan speed control knob (D). These velocities tend to give clearer smoke streams than any other.

Note: After long periods of testing, the smoke nozzles (J) will acquire a liquid buildup which prevents the smoke from exiting them. The machine must be shut down and the smoke nozzles (J) cleaned with pipe cleaners before further operation.

holes allow the test model to be mounted in the smoke chamber. The mount pegs do not interfere with the pattern of the smoke streams. Tenth, place the test model into the smoke chamber and test it. See Figure A-3 for operation instructions.

The materials utilized in the test models are selected for specific purposes. The mat board used for the base must be black in order to show up the kerosene smoke streams, which are white. Clear or black Plexiglas may also be used, but the mat board is more economical, easier to draw on, and as effective as Plexiglas for a base. The walls, or vertical pieces, need to be clear Plexiglas or any other clear, stiff material. The white kerosene smoke is clarified by electric lamps located at the top and bottom of the working section of the smoke chamber. Therefore, the vertical pieces must be clear to allow the light to pass through them and stiff to resist the airflow. The Plexiglas pieces are easily seen even though they are clear; the cut edges are white. Plastic model cement gives good adhesion to the mat board and between two pieces of Plexiglas. The remodeling of a test model is easily accomplished when plastic model cement is used.

During the construction phase of the test models, detailed precision is not important, but accuracy is necessary. The models must be relatively correct so the general air movement patterns are valid. The test models and test equipment are not capable of minute studies of details, such as louver doors, as in a wind tunnel. These issues are important to remember when building the test models for a smoke airflow chamber.

Explanatory Notes

The results of the actual tests conducted should be reviewed with the following conditions in mind.

The studies of the test models are relatively two-dimensional. Therefore, the residential designs are examined in both plan and section, and the data collected are interpreted into three-dimensional information.

The data derived from the smoke chamber are generally qualitative in nature. Quantitative information is not obtainable, since the equipment is not set up for measuring such data. However, the characteristics of the air movement patterns provide insight into the basic velocity of specific airstreams when compared to other airstreams. The relative velocity of the airflow is indicated by the proximity of the smoke streams. When they are far apart, the velocity is low and the air pressure is high; when they are close together, the velocity is high and the air pressure is low.

When no obstructions are located in the smoke chamber, the smoke streams travel through the chamber in uniform and parallel lines. Such laminar flow indicates the absence of air turbulence.

When an obstruction, a model for example, is placed into the smoke chamber, the patterns of the smoke streams are disrupted. The obstruction creates turbulence within the chamber by reducing the initial airflow velocity and redirecting the path of the airflow. The smoke streams may be concentrated and guided into groups of thin streams, or they may be broadened and dissipated into thin air.

Within the immediate vicinity of the test model, the smoke streams are deflected around the model which leaves dark areas between it and closest streams. Areas located on the front face of the model, the side that faces into the airflow, indicate regions of high pressure; areas located on the sides and back face of the model indicate regions of low pressure.

The evaluation of the test models occurs in a two-dimensional format. The illustrations in Chapters 12 and 13 demonstrate the basic principles of air movement which are graphically and qualitatively valid.

Bibliography

The American Institute of Architects, *AIA Energy Notebook,* vol. 1, The American Institute of Architects, Washington, D.C., 1975.

American Society of Heating, Refrigerating, and Air-Conditioning Engineers, Inc., *ASHRAE Handbook—1985 Fundamentals,* American Society of Heating, Refrigerating, and Air-Conditioning Engineers, Inc., New York, 1985.

Gary O. Robinette and Charles McClennon, *Landscape Planning for Energy Conservation,* Van Nostrand Reinhold, New York, 1983.

The Architects Collaborative, "Terraced Pods Invite Daylight and Breezes," *Architectural Record,* vol. 169, Mid-August 1981, pp. 52–57.

Aren, Edward A., and Philip B. Williams, "The Effect of Wind on Energy Consumption in Buildings," *Energy and Buildings,* vol. 1, May 1977, pp. 77–84.

Aronin, Jeffrey Ellis, *Climate and Architecture,* Van Nostrand Reinhold, New York, 1953.

Bahadori, Mehdi N., "Passive Cooling Systems in Iranian Architecture," *Scientific American,* February 1978, pp. 144–152.

Banham, Reyner, *The Architecture of the Well-Tempered Environment,* The University of Chicago Press, Chicago, 1969.

Barbosa, Susan, "Put Nature to Work Saving Energy," *The Ledger,* Lakeland, Fla., vol. 14, February 1981, C, p.1.

Barr, Tom, "Cool as an Onion," *New Shelter,* July/August 1982, pp. 29–31.

Bazjanac, Vladimir, "Energy Analysis: Crystal Cathedral, Garden Grove, California," *Progressive Architecture,* Penton/IPC, vol. 61, December 1980, pp. 84–85.

Bowker, Michael, and Jim Finley, "Mind Your Climate," *New Shelter,* July/August 1982, pp. 23, 24.

Brooks, C. E. P., *Climate in Everyday Life,* Philosophical Library, New York, 1951.

Brubaker, C. William, "Energy Conservation through Rational Architecture and Planning," *Building Systems Design,* vol. 73, June/July 1976, pp. 10–20.

The Bureau of Research, *Houses and Climate: An Energy Perspective for Florida Builders,* The Governor's Energy Office, Tallahassee, Fla., 1979.

Caudill, William W., Sherman E. Crites, and Elmer G. Smith, *Some General Considerations in the Natural Ventilation of Buildings,* Research Report No. 22, Texas Engineering Experiment Station, College Station, Tex., 1951.

―――― Frank D. Lawyer, and Thomas A. Bullock, *A Bucket of Oil,* Cahners Books, Boston, 1974.

―――― and Bob H. Reed, *Geometry of Classrooms as Related to Natural Lighting and Natural Ventilation,* Research Report No. 36, Texas Engineering Experiment Station, College Station, Tex., 1952.

Chandra, Subrato, Philip W. Fairey, III, Arthur B. Bowen, Jack E. Cermak, and Jon A. Peterka, *Passive Cooling by Natural Ventilation: A Review and Research Plan,* Florida Solar Energy Center, Cape Canaveral, Fla., 1981.

Chermayeff, Serge, and Christopher Alexander, *Community and Privacy,* Doubleday, Garden City, N.Y., 1963.

Ching, Francis D. K., *Architecture: Form, Space and Order,* Van Nostrand Reinhold, New York, 1979.

Clark, Wilson, *Energy for Survival,* Anchor Books, Garden City, N.Y., 1975.

"Coastal Cottage Catches the Breeze," *Southern Living,* June 1982, p. 160.

Crump, Ralph W., "Games that Buildings Play with Winds," *AIA Journal,* vol. 61, March 1974, pp. 38–40.

Department of Economics and Social Affairs, *Climate and House Design*, United Nations, New York, 1971.

Department of Housing and Community Development, *Energy Design Manual*, Department of Housing and Community Development, Sacramento, Calif., 1975.

Dubin, Fred S., and Chalmers G. Long, Jr., *Energy Conservation Standards*, McGraw-Hill, New York, 1978.

Evans, Benjamin H., *Natural Air Flow around Buildings, Research Report No. 59*, Texas Engineering Experiment Station, College Station, Tex., 1957.

Fiebach, David R., *Housing Option to Reduce Fossil Fuel Dependency*, master's thesis, University of Florida, Gainesville, Fla., 1978.

Fischer, Robert E., "The Crystal Cathedral: Embodiment of Light and Nature," *Architectural Record*, vol. 168, November 1980, pp. 77–85.

Flower, Robert G., and Frederic S. Langa, "Fresh Air Without Frostbite," *New Shelter*, January 1984, pp. 58–60, 62–64, 66–67.

Fry, Maxwell, and Jane Drew, *Tropical Architecture in the Dry and Humid Zones*, Van Nostrand Reinhold, New York, 1964.

Geiger, Rudolf, *The Climate Near the Ground*, Harvard University Press, Cambridge, Mass., 1959.

Givoni, B., *Man, Climate, and Architecture*, Applied Science, Ltd., London, 1976.

Goldstein, Barbara, "New Crystal Palace," *Progressive Architecture*, vol. 61, December 1980, pp. 76–82.

Harding, Louis Allen, and Arthur Cutts Willard, *Heating, Ventilating and Air Conditioning*, Wiley, New York, 1937.

Higson, James D., *Building and Remodeling for Energy Savings*, Craftsman Book, Solana Beach, Calif., 1977.

Hill, Burt, and Associates, *Planning and Building the Minimum Energy Dwelling*, Craftsman Book, Solana Beach, Calif., 1977.

Holleman, Theo R., *Air Flow through Conventional Window Openings, Research Report No. 33*, Texas Engineering Experiment Station, College Station, Tex., 1951.

Jacobs, Madeleine, "Summer Tips for Saving Energy and Money," *Dimensions/National Bureau of Standards*, vol. 61, July 1977, pp. 8–14.

Jarmul, Seymour, *The Architect's Guide to Energy Conservation*, McGraw-Hill, New York, 1980.

Kals, W. S., *The Riddle of the Winds*, Doubleday, Garden City, N.Y., 1977.

Kinzey, Bertram Y., Jr., and Howard M. Sharp, *Environmental Technologies in Architecture*, Prentice-Hall, Englewood Cliffs, N.J., 1963.

Lacy, R. E., *Climate and Building in Britain*, Building Research Establishment, London, 1977.

Lafavore, Michael, "A Healthy House Tour," *New Shelter*, May/June, 1982, pp. 29–31.

——— "One Sure Cure," *New Shelter*, May/June 1982, pp. 26, 27.

——— "Something's in the Air," *New Shelter*, May/June 1982, pp. 20–25.

Langa, Frederic S., "A Solar Air Conditioner," *New Shelter*, July/August 1983, pp. 18, 20.

——— "Cooling without Kilowatts," *New Shelter*, July/August 1982, p. 17.

——— "Many Ways to Cut It," *New Shelter*, July/August 1982, pp. 18–22.

Lesiuk, Stephen, "Architectural and Environmental Horticulture: An Investigation into the Use of Vegetation for Energy Conservation," *Design: Research, Theory, and Application*, vol. 10, Environmental Design Research Association, Washington, D.C., 1979.

Lindsely, E. F., "Solar Air Conditioners," *Popular Science*, July 1984, pp. 64–66, 99.

Lynch, Kevin, *Site Planning*, The M.I.T. Press, Cambridge, Mass., 1971.

Mehta, R. D., and P. Bradshaw, "Design Rules for Small Low Speed Wind Tunnels," *The Aeronautical Journal*, vol. 83, November 1979, pp. 443–449.

Mitchell, Chuck, "One Man's Bevy of Heat Beaters," *New Shelter*, July/August 1982, pp. 25–28.

Morton, David, "'Dog Trot' House," *Progressive Architecture*, vol. 62, June 1981, pp. 86–89.

——— "The Elements and Form," *Progressive Architecture*, vol. 62, April 1981, pp. 108–113.

National Oceanic and Atmospheric Administration and National Climatic Data Center, *Local Climatological Data, Monthly Summary,* U.S. Department of Commerce, National Climatic Data Center, Asheville, N.C., 1984.

Nevrala, D. J., and D. W. Etheridge, "Natural Ventilation in Well-Insulated Houses," *Energy Conservation in Heating, Cooling, and Ventilating Buildings,* vol. 1, Hemisphere Publishing, Hemisphere, Wash., 1978.

Odum, Howard T., and Elisabeth C. Odum, *Energy Basis for Man and Nature,* 2d ed., McGraw-Hill, New York, 1981.

O'Hare, Michael, and Richard E. Kronauer, "Fence Designs to Keep Wind from Being a Nuisance," *Architectural Record,* vol. 146, July 1969, pp. 151–156.

Olgyay, Aladar, and Victor Olgyay, *Solar Control and Shading Devices,* Princeton University Press, Princeton, N.J., 1957.

Olgyay, Victor, *Design with Climate,* Princeton University Press, Princeton, N.J., 1963.

"One-Story Contemporary Houses," *Houses and Plans,* Hearst Corp., 1982, pp. 60–65.

Ortner, Everett H., "Natural Ventilation," *Windows and Glass in the Exterior of Buildings,* Building Research Institute, Washington, D.C., 1957.

"Replace a Ceiling Light with a Fan," *Southern Living,* July 1982, pp. 110–113.

Rush, Richard, "Hotsification," *Progressive Architecture,* vol. 62, April 1981, pp. 114–117.

Scheller, William G., *Energy-Saving Home Improvements,* Howard W. Sams, Indianapolis, Ind., 1979.

Schmertz, Mildred F., "Bratti House, New Canaan, Connecticut," *Architectural Record,* vol. 169, Mid-May 1981, pp. 94–97.

"Shutters that Also Shade," *Southern Living,* June 1982, p. 147.

Sizemore, Michael M., Henry O. Clark, and William S. Ostrander, *Energy Planning for Buildings,* Michael M. Sizemore and the American Institute of Architects, Altanta, Ga., 1979.

Skipper, Suzanna, *Cracker Houses: Low Energy Comfort,* master's thesis, University of Florida, Gainesville, Fla., 1980.

Smay, V. Elaine, "Heat-Saving Vents—Are They the Solution to Indoor Pollution?" *Popular Science,* January 1983, pp. 78–81.

Smith, Elmer, G., *The Feasibility of Using Models for Predetermining Natural Ventilation, Research Report No. 26,* Texas Engineering Experiment Station, College Station, Tex., 1951.

——— Bob H. Reed, and H. Darwin Hodges, *The Measurement of Low Air Speeds by the Use of Titanium Tetrachloride, Research Report No. 25,* Texas Engineering Experiment Station, College Station, Tex., 1951.

Smolen, Marguerite, "The Truth about Cool Tubes," *New Shelter,* August 1984, pp. 57–59.

Smyser, Carol A., and the Editors of Rodale Press Books, *Nature's Design,* Rodale Press, Emmaus, Pa., 1982.

Stains, Larry, "Starting Thoughts—The Most Energy-Efficient Home in America," *New Shelter,* February 1983, pp. 8–12.

Stepler, Richard, "Passive-Solar Kit Homes," *Popular Science,* May 1981, pp. 104–106.

Texas A&M University, *Environmental Criteria: MR Preschool Day Care Facilities,* Department of Health, Education, and Welfare, Washington, D.C., n.d.

Thomson, William A. R., *A Change of Air,* Scribner's, New York, 1979.

Trumper, H., and H. Bley, "Ventilation of Flats," *Energy Conservation in Heating, Cooling, and Ventilating Buildings,* vol. 1, Hemisphere Publishing, Hemisphere, Wash., 1978.

"20% Cooler without Air Conditioning," *House and Home,* February 1955, pp. 140–143.

United States Department of Energy, *Tips for Energy Savers,* United States Department of Energy, Washington, D.C., 1978.

Wagner, Walter F., Jr., and Robert E. Fischer, "Round Table: A Realistic Look at 'the Passive Approach'—Using Natural Means to Conserve Energy," *Architectural Record,* vol. 168, Mid-August 1980, pp. 92–100.

Warren, P. R., "Ventilation through Openings on One Wall Only," *Energy Conservation in Heating, Cooling, and Ventilating Buildings,* vol. 1, Hemisphere Publishing, Hemisphere, Wash., 1978.

Watson, Donald, *Energy Conservation through Building Design*, McGraw-Hill, New York, 1979.

———— "The Energy Within the Space Within," *Progressive Architecture, Penton/IPC*, vol. 63, July 1982, pp. 97–102.

"Window Design Report Gives Energy Strategies," *AIA Journal*, vol. 66, September 1977, pp. 92–94.

Wise, A. F. E., "Ventilation of Buildings: A Review with Emphasis on the Effects of Wind," *Energy Conservation in Heating, Cooling, and Ventilating Buildings*, vol. 1, Hemisphere Publishing, Hemisphere, Wash., 1978.

Index

ABOUT THE AUTHOR

Terry S. Boutet, AIA, received both his bachelor of design and his master of architecture degrees from the University of Florida at Gainesville. While a graduate student there, he conducted the research for this book and experimented with the smoke airflow chamber. Mr. Boutet is also a landscape artist who works in acrylics and oils.